THE FORGOTTEN

THE FORGOTTEN FLAG
THE INSPIRATION

THE FORGOTTEN WAR
OUR HEROES' STORIES

THE FORGOTTEN VICTORY
A COUNTRY'S SALVATION

~Volume One~

WILLIAM A. CUMMINS

CAI Publishing
Port Orange
Florida

THE WORLD OF BOOKS

The world of books
is the most remarkable creation of man.
Nothing else that he builds ever lasts —
monuments fall;
nations perish;
civilizations grow old and die out
and after an era of darkness
new races build others.

But in the world of books are volumes
that have seen this happen again and again
and yet live on;
still young,
still as fresh as the day they were written;
still telling men's hearts
of the hearts of men centuries dead.

CLARENCE DAY

American Humorist,
Essayist, Biographer and Writer

What Others Are Saying About This Book

"The Marine Corps photo on the cover of this book captures a rare moment in time when Private First Class Luther R. Leguire planted the Stars and Stripes atop the American Ambassador's residence. His actions symbolized the capture of the Korean Capitol by the United Nations Forces in 1950. Later wounded and left for dead, he survived to fulfill a spiritual commitment made during battle. This inspirational book preserves his and many more self-told stories of the historical epic war called, THE FORGOTTEN."

GENERAL AL GRAY, USMC (RET.)
29th Commandant of the Marine Corps (Four Star)

"Past generations maintained history by passing stories from memory to all who would listen and do likewise. History was more than stories — it was personal. Although we have better ways to record facts today with print, video, and audio, these have left many disinterested in the past.

The past comes alive again in this splendid new book, THE FORGOTTEN. It captures the personal part of the Korean War that is lost in so many efforts to explain or understand that conflict. Make no mistake — if you want to know what Truman, Stalin, Mao, or Kim Il-Sung were thinking — this is NOT your book.

But, if you believe that history is more than just academic and is instead a mosaic of the human experience, then prepare to see, understand, and remember Korea differently and more personally than before. This is history, not as we were taught, but as lived by those who were there."

LIEUTENANT COLONEL WM. SCOTT PHILLIPS,
MILITARY INTELLIGENCE, US Military Academy; Class of 1984

"It is difficult to imagine the anguish, the determination and yes even the fear, in the heart of a man driven by enemy fire into a fox hole. The man who just received an order to take a hill that was lost to the enemy has those same feelings. A fighter pilot about to roll in on a well defended target, for a split second, may have similar thoughts.

What a wonderful objective it is to capture the thoughts and actions of these men in combat. William A. Cummins has met this challenge with his remarkable book entitled, THE FORGOTTEN — a series of stories by Korean War veterans depicting their combat experiences in their very own words. Experiences from a war that must never be forgotten."

MAJOR GENERAL FREDERICK C. 'Boots' BLESSE,
USAF (RET) Korean War Fighter Pilot Ace

THE FORGOTTEN
The Forgotten Flag — The Forgotten War — The Forgotten Victory

© 2009 by William A. Cummins

ISBN Print Edition: 978-0-9787766-1-9
ISBN PDF eBook Edition: 978-0-9787766-2-6

Published by CAI Publishing
807 Black Duck Drive, Suite A
Port Orange, FL 32127-4726, USA
Web Site: *http://www.caipublishing.net*
Email: *info@caipublishing.net*
Office: 386.383.5198

All rights reserved. No part of this book may be reproduced or transmitted in any form or by any means, electronic or mechanical, including photo-copying, recording or by any information storage and retrieval system without written permission from the author, except for the inclusion of brief quotations in a review.

Cover Photo: Seoul, Korea, September 26, 1950. Amid sniper fire a dirty-faced Marine Private First Class, Luther R. Leguire of Dayton, Ohio, planted the Stars and Stripes on the porch roof of the American Ambassador's residence at 3:37 P.M. United Press Staff Correspondent, R. Vermillion.

Photo Credit — Hqtrs. No. A3386 — Sgt. John Babyak, USMC
Printed in the United States of America
Library of Congress Cataloging-In-Publication Data

Design and layout by Martha Nichols/aMuse Productions®
Cover design by John Morris-Reihl/Art and Technology

FOREWORD

By Historian and Former Six Term
Ohio Congressman, Bob McEwen

"THE FORGOTTEN." HOW CAN ANY SOLDIER BE FORGOTTEN? FOR over half a century we have asked that question. How can one forget the 36,940 soldiers who died or the 92,134 wounded, during those three hellish years in Korea lasting from June 25, 1950 to July 27, 1953?

There are 8,176 still missing in action so again, we must ask, "Why was this war forgotten?"

Is it because the war was never formally ended or that this conflict is only held in check by a tense DMZ ceasefire agreement? Is that the reason thousands of U.S. troops have been stationed in Korea for over 55 years?

Sadly the answer is, "Yes." The Korean War will not end until North Korea ceases its hostile threats that continue even today against South Korea and its friends.

Without making a fuss, the American combat troops came home and quietly melded into the fabric of our society. There was no media outpouring about where they had been, what they had seen, or what they had endured. There was Silence Only Silence.

How could the January 1, 1951 *Time Magazine* cover entitled 'G.I. Joe — Man of the Year' fade into obscurity? It is now six decades later, and those brave men and women are passing from us in very large numbers, taking their stories with them. Soon lost to antiquity!

Questions abound. What happened to them during their deployments in Korea and what has happened to them since the war? Thankfully, we have several of their stories in this thoughtfully written volume by William A. Cummins.

These are untold stories, written in their own words, of personal military experiences. Some of their narratives are funny while others are heartbreaking and sad. Many of the men simply wanted their stories to be remembered by their grandchildren.

This is an inspiring book of heroes that not only makes the reader feel a part of their combat but also insures that we will not forget their great sacrifices. One of those heroes is my personal friend, Luther R. Leguire, whose story as a young Marine PFC is highlighted in the first section of the book. He is also seen on the book's cover.

You will appreciate with me his climb onto the porch roof of the American Ambassador's residence in Seoul, Korea at the War's beginning. How, amid sniper fire, he removed the North Korean flag and planted the first Stars and Stripes in the Capital City. And how, a few days later, he was left for dead after an enemy ambush that killed all of his Marine buddies.

Those life-changing battlefield experiences eventually brought Luther Leguire to Lake City, Florida where, in 1962, he established a church to fulfill his promise to God. He has become an outstanding leader who now touches the life of each person he meets.

You will share the image of the bombs falling from the underbelly of our American bombers that blasted the enemy ground positions prior to the United States Marines invasion at Inchon. This book tells the story of those who survived the horrific winters on the front lines and the battle that split our forces at the Chosin Reservoir.

Revealed is the thrill of a young naval photographer in the Gunner's seat of an AD Raider Aircraft as he is 'flung out into space' (the first time) from the Carrier's flight deck. He is flooded with FEAR as he approaches his ship to land. With eyes closed, 'flaps' down, and 'tail hook' dragging he hits the deck with a big 'BANG' — Mission Accomplished!

The readers of this book will not forget the huge Victory for the people of South Korea. Established after WWII, it has developed a successful democracy and is now the economic envy of developing countries everywhere. South Korea is one of the wealthiest of tigers in all of Asia.

Following the war, South Korea became one of America's most reliable friends, and its people have been forever grateful to America for the sacrifices made by our men and women in the Forgotten Korean War.

"Together, We Honor, Salute, And Say Thank You To Our Heroes With This Book!"

Bob McEwen

DEDICATION

KOREAN WAR VETERANS ARE PARENTS AND GRANDPARENTS today. Most are more than 75 years old and have lived through at least six wars and 13 US Presidents. They may be the last generation to have seen, or even remember, anything like the Korean War and World War II.

This book is a testimony that AMERICAN patriotism keeps AMERICA great. It is a witness that the American dream still lives and will never die in the hearts of people who love and cherish freedom everywhere.

Initially, this book is dedicated to the courageous men and women who left their homes and fought against Communism in Korea, facing hardship and even death for the sake of VICTORY and FREEDOM.

Next, this book is dedicated to the dauntless men and women who voluntarily serve in the armed forces in order to protect their families from terror and tyranny. These soldiers stand ready to defend our land wherever they are needed.

Lastly, this book is dedicated to all who patiently wait at home for their gallant loved ones. Many do not return, and this loss overloads their hearts with sadness and emptiness.

Let this book remind us all that Freedom comes at a great price for every generation. It is inescapable that the price of Freedom demands the willingness to sacrifice the lives of our young warriors who protect and defend it, not only for ourselves but for future generations.

> "The legacy of heroes is the memory of a great name and the inheritance of a great example."
>
> *Benjamin Disraeli*

EPIGRAPH

HOW CAN YOU FORGET A WAR? HOW CAN YOU LOSE SIGHT OF what happens during wartime? How can you forget all the wars that were fought to defend the freedom and peace of people throughout the World?

What happened to those gallant warriors who risked their lives, families, and fortunes to help oppressed peoples in their battles against tyranny? What happened to all those precious souls when they returned home after witnessing the agony of battle?

Why is it so hard to remember the events that lead up to conflict where war remains as the only solution? Is our mind programmed to forget pain lest it overcome our emotions? Do we believe by forgetting war, we can free ourselves of the responsibility to honor our warriors for their sacrifices?

This book contains the untold story of the hoisting of the stars and stripes in a country that bravely rose up against its oppressors, and went on to produce one of the greatest postwar victories in contemporary history.

It presents many personal stories of soldiers as they left home responding to their call to arms. Men now in their senior years, who realize that time keeps marching on. Their self told war time stories reveal to the world how important it is to stand up, and do the right thing when called into battle.

THIS BOOK IS A TESTIMONY TO THEIR DEDICATION.

"We must learn to regard people less in light of what they do or omit to do, and more in the light of what they suffer."
Dietrich Bonhoeffer

ACKNOWLEDGMENTS

Inspiration

I am especially grateful that Pastor Luther R. Leguire shared how his whole life was impacted and changed by the bloody battlefield experiences endured during the Korean War while serving as a young soldier in the US Marine Corps.

Vision

Invaluable was the vision set forth by Retired Navy Chief Petty Officer, Sam Kemp, who first conceived the idea for this book, and started it on a steady course to completion.

Korean Veterans

Like most WWII Veterans before them, they simply tried to put the war out of their minds by not talking about it, even to each other. It is a great honor to include so many of their self told stories, which make this book so fulfilling.

Support

My special and lasting appreciation goes to Peggy Painter for her careful editing, seamless phrasing, and helpful commentary offered during the completion of the book. Most importantly, I want to express gratitude to my wife, Ann, for her boundless patience, encouragement, faith, and assistance during the researching and writing of this book.

DISCLAIMER

THIS BOOK IS DESIGNED TO PROVIDE PERSONAL INSIGHT, AND historical information on the lasting effects of the Korean War. It is sold with the understanding that the publisher and author are not attempting to redefine history. If legal or other expert assistance is desired, the services of a competent historian should be sought.

It is not the purpose of this book to reprint information otherwise available to authors and other creative people, but to complement, amplify, and supplement other texts. You are invited to pursue your interests in libraries, book stores, and the Internet. Resources are abundantly available to anyone interested in the events presented herein.

Historical information was primarily gathered through Wikipedia. Korean War causality information was obtained from the Office of Secretary of Defense, Washington Headquarters Services.

Veteran stories remained untouched except for punctuation, and spelling accuracy. Our desire was that all the veterans would convey their own style of story telling, and thereby make their stories more enjoyable to their families and to all the new friends who get to know them through this book.

Every effort has been made to make this book as accurate as possible. However, there may be mistakes, both typographical and in content. Therefore the text should be used only as a general guide to the historical events that took place surrounding the Korean War.

The purpose of this book is to educate, inspire, and entertain. The author and publisher shall have neither liability nor responsibility to any person, directly or indirectly, for any problems or troubles alleged to have been caused by the information in this book.

If you do not wish to be bound by the above, you may return this book to the publisher for a full refund.

TABLE OF CONTENTS

Introduction xvii

Section I — THE FORGOTTEN FLAG *The Inspiration* 1
  Chapter 1. Who Is Luther R. Leguire 3
  Chapter 2. Sharecropper - World War II 7
  Chapter 3. Leaving Home - The Talladega Trip 13
  Chapter 4. Planting Old Glory - United Press Article 17
  Chapter 5. All Hell Broke Loose — I Remember Praying 21
  Chapter 6. The Faith Journey — God Moved On Me 29
  Chapter 7. Youth Boot Camp — Leguire's Legacy 41

Section II — THE FORGOTTEN WAR *Our Heroes' Stories* ... 47
  Chapter 1. The Inchon Invasion 49
  Chapter 2. Chinese Intervention 53
  Chapter 3. Stalemate at the DMZ 57
  Chapter 4. Our Heroes' Stories In Their Own Words 63
    Rogers, Dr. Carl T. — Lawrenceville, GA 65
    Cowan, William A. — Saginaw, TX 72
    Baumann, Donald — Cocoa Beach, FL 77
    Grogan Jr., Stanley — Pinole, CA 81
    Walker, Bryan 'Jerry' — Kansas City, MO 84
    Appenzeller, Charles — Haines City, FL 88
    Brennen, William O. — Larkspur, CA 90
    Hanen, Dale L. — Brunswick, GA 97
    Groves, Harold L. — Jackson, TN 100
    Kennedy, John Rick — Port Orange, FL 102
    Street, Phil — Jonesboro, IN 109
    McGuire, Robert — Daytona Beach, FL 112
    Palese, Dr. John A. — Milwaukee, WI 114
    Furuichi, Fred — Fairfield, CA 116
    Rabasco, Anthony — Scarsdale, NY 124
    Raber, Robert — Sequim, WA 127

Introduction 131
   Davis Sr., Raymond A. 132
   Jackson, Andrew B. — Fredericksburg, VA 133
   Marbaugh, Roland A. — Conyers, GA 135
   Hall, William — Atlanta, GA 137
   Sabel, Norman — Stone Mountain, GA 140
   Norton, Dewey — Milledgeville, GA 142
   Landers, Alvin — Warner Robbins, GA 144
   Sofo, Faust — Staten Island, NY 146
   Sweeney, Bob — Waycross, GA 148
      Poems (2): Bob Darling 150
      Poem: Frank Youngman 152
   Olwine, Robert T. — Nokomis, FL 153
   Ruffing, Leo G. — Portsmouth, VA 166
   Ward, Leon B. — Palm Coast, FL 172
   Worrill, Frederick W. — Crawfordville, FL 175
   Blesse, Frederick 'Boots" — Melbourne, FL 180
   Wolfe, Ben "Doc" 186
   Wilson, Robert C. — Clio, MI 193
   Cummins Brothers — Lewistown, OH 204
Wrap Up 205

Section III —
THE FORGOTTEN VICTORY A Country's Salvation 207
   Chapter 1. Korean War — Global Perspective 209
   Chapter 2. Korean War — Historical Review 213
   Chapter 3. Korea Today — The Victory 221
   Chapter 4. Korean War Veterans Memorial 233

Appendix Wanted: War Veteran Stories 237

Glossary 239

Colophon 240

Quick Order Form 241

INTRODUCTION

SOUTH KOREA'S ECONOMY TODAY IS ONE THE WORLD'S MOST stunning achievements. This achievement, however, was preceded by a bloody three year war which claimed 36,940 American lives; left another 92,134 wounded, and counts 8,176 still missing in action.

Why North and South Korea required intervention by the United Nation's has been the subject of much historical debate and discussion. Each country lost over one million lives during this struggle, and the peninsula was devastated.

Equally important is the realization that for the first time our country tried to forget the consequences of not winning a war. The Korean War ended in a stalemate that continues to this day, secured only by the continued presence of armed troops stationed on both sides of the 38th Parallel.

There have been more than 40,000 documented cease fire violations and hostile acts reported since the war ended in 1953. Forgotten also are more than 1,200 US and 2,300 Republic Of Korea military personnel who have given their lives defending South Korea during that period.

Our battle weary troops who fought the vicious North Korean and Chinese enemy, returned home to a country that ignored their heroic efforts, and the terrible sacrifices they endured.

It is the desire of this writer that the following work becomes a bridge over our unjust behavior toward these brave men and women and to recognize what their humble sacrifices have produced throughout the free world.

THE FORGOTTEN FLAG
THE INSPIRATION

We begin by honoring the heroic exploit of a dirty-faced 19 year old PFC Marine who, amid sniper fire, planted the stars and stripes on top of the roof of the US Ambassador's residence in Seoul, Korea on September 26, 1950.

Ambushed a few days later and left for dead, Luther R. Leguire watched as everyone in his outfit was killed by the invading Chinese Communist Army. It was on this battlefield that his heart was changed.

You will discover his grateful heart as he travels from share-cropping as a youth, to deadly combat as a Marine, to soul-cropping as the Pastor of his own church.

THE FORGOTTEN WAR
OUR HEROES' STORIES

Many of our Veterans have already died without sharing their experiences. Following a general outline of the Korean War, you will discover for the first time many untold personal stories from our Veterans who actively fought its battles.

A defeated North Korean Army was suddenly reinforced by millions of Communist Chinese forces pouring across the border to stop the advancing UN Forces in their tracks.

During the height of this war, on April 11, 1951, General Douglas MacArthur was relieved of his command in the Far East by U.S. President Harry S. Truman, who had ordered the dropping of the first atomic bomb only six years earlier.

Never underestimate the impact of this war in stopping the advancement of Communism throughout the world.

THE FORGOTTEN VICTORY
A COUNTRY'S SALVATION

Finally revealed is the economic triumph of the South Korean people. With their land laid waste by war, they had to rebuild their economy and infrastructure from the ground up after the armistice ended the fighting in 1953.

Newly freed from their northern Communist oppressors, and backed by the US, the South Koreans chose hard work, western style capitalism, and a democratic government to achieve one of the strongest economies in the world.

A strategic alliance developed between South Korea and the US after the war. It has participated in most of the military conflicts the US has been involved in since 1953.

South Korea has never forgotten and will be forever grateful to America for the sacrifices of its men and women who gave them freedom from their Communist neighbors.

THE FORGOTTEN FLAG

The Inspiration

Marine PFC Luther R. Leguire planting the American flag in Seoul, Korea; September 26, 1950.

"I like Marines, because being a Marine is serious business. We're not a social club or a fraternal organization and we don't pretend to be one. We're a brotherhood of "warriors"... nothing more, nothing less, pure and simple. We are in the ass-kicking business, and unfortunately, these days business is good."

Colonel James M. Lowe, Commander,
Marine Corps Base Quantico, 2004

"Some people spend an entire lifetime wondering if they made a difference in the world. But, the Marines don't have that problem."

Ronald Reagan

Chapter One

WHO IS LUTHER R. LEGUIRE?

LUTHER R. LEGUIRE WAS BORN DECEMBER 18, 1930 IN DAYTON, Ohio. He grew up in a sharecropper's family of nine children. Luther quit school in the ninth grade, and at age 15 left his home behind. He liked being on his own, and at age 17, joined the United States Marine Corps on January 26, 1948.

After attending 13 weeks of boot camp at the Marine Corps Recruit Depot in Parris Island, South Carolina, Private Leguire completed a Mediterranean Cruise in 1949. Then on September 15, 1950 he went ashore in the initial Marine Landing at Inchon, South Korea.

On September 26, 1950 during heavy fighting for the capture of the city of Seoul with E Co 2nd Bn 1st Marine Regiment, PFC Leguire climbed atop the porch roof of the American Ambassador's residence. Amid sniper fire, he removed the North Korean flag that his unit had already shot full of holes and he replaced it with the American Flag.

About two months later, during an enemy ambush on November 7, 1950, near the North Korean City of Koto-ri, PFC Leguire was wounded and left for dead. Lying wounded and alone Leguire avowed, "God, If you'll let me live and get home, I'll serve you the rest of my life!" He was the only man in his unit to survive.

He was eventually transferred to the Naval Hospital in Jacksonville, Florida to recover from his wounds. He was awarded not only the Purple Heart, but the Presidential Unit Citation, The Korean Presidential Unit Citation, and Korean Service Medal as well. On November 30, 1951, Leguire officially retired from the Marines with the rank of Corporal.

Leguire remained in Jacksonville where he put the war out of his mind, married, and started a family. Heeding the call to preach, he moved his family to Lake City, Florida in 1962 where he founded the First Apostolic Church of Lake City, Inc. and remains their beloved Pastor today.

In October 2007, Pastor Leguire was contacted by a Marine Colonel and asked if he was the same Leguire that raised the flag over the American Ambassador's residence in Seoul during the beginning of the Korean War. When the Colonel verified Leguire was the man, he invited him to South Korea for the Marine Corps Birthday Celebration the coming November.

The night before the Ball, at a special dinner, Leguire and his associate, Deanna King, discovered that the entire celebration was centered on the photograph of Leguire planting the stars and stripes atop the porch roof of the American Ambassador's Residence in Seoul.

The next evening when the General rose to deliver his remarks, he asked the audience to look at the Marine Corps programs on their tables. He pointed to the cover picture showing a man planting the US flag and said, "That man, revealed in the photograph, is with us this evening."

Pastor Leguire was asked to stand and be recognized. He was greeted with an ovation that left him truly surprised and humbled. It seems as though, unknown to this modest pastor, this photograph was quite famous in Korean War Veteran circles.

Retired four-star General and 29th Commandant of the Marine Corps, General Alfred M. Gray, who had never met Leguire, traveled to Korea for the celebration after hearing that Luther would be present. General Gray is known as a "Warrior Commandant." Upon Leguire's introduction that evening, General Gray moved to Corporal Leguire's table and gave him a big hug! Private First Class Luther R. Leguire was unaware that he had been a local hero to the South Korean people for more than 57 years!

Later, Leguire showed his hosts the remains of the North Korean flag he had removed from the Ambassador's residence. He said he had placed it in a paper bag and all but forgotten about it during those 57 intervening years.

The next day he visited several local areas in Seoul. When the older Korean people were told who he was, they immediately bowed politely and told him how much they appreciated what he had done for them.

It seems that historically, the raising of the American flag over Seoul means as much to the South Koreans as the raising of the US flag at Iwo Jima does to the Marine Corps.

We all salute you, retired Corporal Luther R. Leguire!

"To all great and courageous soldiers — I humbly salute you. I also salute this country's great and everlasting "Stars and Stripes," and may no power arise to take it down."

Corporal Luther R. Leguire,
USMC Retired

Top left: Planting the Stars and Stripes; PFC Luther R. Leguire, 19 year old Marine; Top right: North Korean Banner; Bottom left and right: Leguire at Korean Veterans Memorials — Seoul, South Korea in 2007 and Washington, DC in 2008.

Chapter Two

SHARECROPPER — WORLD WAR II

THERE MAY HAVE BEEN SOMETHING SPECIAL ABOUT THIS BABY, but arriving wet, cold, and crying did not seem like a good idea to the infant. He gulped air for the first time and suddenly realized he had just emerged into a strange new world.

Luther Robert Leguire was the youngest child of Clyde and Nora Leguire. Their oldest child, Punky, died at age six from diphtheria before Luther was born. His two older brothers were Glen Eugene and Harry Wallace Leguire.

Luther wasn't the first baby boy born in Dayton, Ohio during the Great Depression and he certainly wasn't the last. However, something very special happened to Luther R. Leguire on his way to manhood and this is his story.

Luther (hereinafter called Leguire) was born during the darkest three years of the Great Economic Depression. It was America's worst encounter with deflation when prices fell about 10 percent every year. From 1930-1933, financial activity of all kinds including loans, consumer lending, and mortgages, almost entirely ceased.

Massive financial problems, including, bankruptcies, defaults, and bank failures, were endemic. It was truly a spectacular economic train wreck through out the United States and the world. Yet America still had values, ideals, hopes, and the belief that things would improve.

Many people, however, saw life drained of all meaning. Unemployment rose to 25% and the economy shrank by 26% in three years. The forces that initially sent the economy into a depression in 1928 and 1929 were still at work: a stock market boom turned bust, unprecedented levels of consumer debt, and a real estate boom that turned bust. The lowest recorded birth rates in the US occurred during the Great Depression.

HARDTIMES

Although out of work, Leguire's Dad and Mom made him feel welcome as they set their sights on simply staying alive. They soon moved to southern Ohio, to a little town called Franklin Furnace located about 16 miles east of Portsmouth along the North side of the Ohio River.

Times were tough, and work was hard to find. Keeping a roof over their heads and food on the table was a back- breaking job. Leguire's dad was a house painter which meant lots of hard work for very little money. When Leguire was five years old his father suddenly died. Destitute, his mother remarried and over time bore six more children.

Their new life as sharecroppers meant giving half their crop money to the land owner. Sometimes they only rented. To Leguire It seemed they moved about every time the rent was due. The family eventually moved to Harlan, Kentucky looking for work in the coal mines.

Continually seeking work eventually brought them back to several small towns in Ohio. They returned to Franklin Furnace, then to Jasper in 1942, and back to Franklin Furnace in 1946.

Everyone worked to make money for the family. When Leguire was old enough, he made money shining shoes in the local village. Some days he brought home 80 cents which he earned at a nickel a shine. His mother was very proud of him and wrote down in a journal every penny he made. She zealously subtracted every cent in his book and wrote what she bought with it: new shoes, or eggs and bacon, perhaps a small gift. All were written in his journal.

Young Leguire didn't really dwell on being poor because everyone was poor. Wearing hand-me-down clothes with patches on the knees and elbows was the norm. New shoes for school were rare, and folded cardboard stuffed in the bottom of his shoes for makeshift soles was no big deal. He didn't complain; at least he had some shoes to wear.

WORLD WAR II

The Japanese attack of Pearl Harbor on December 7th, 1941, was devastating. The next day, America was in the greatest struggle for freedom the world has ever witnessed. Time seemed to stand still as the reality of a world-wide war slowly dawned.

Leguire was only 11 years old when the nation started to prepare for the magnitude of this threat. The newspapers, radio, and newsreels started blazing out the headlines of eminent German and Japanese attacks on our mainland.

The next four and a half years were dedicated solely to winning the war in Europe and in the Pacific. Every able bodied man below age 40 either volunteered or prepared to be drafted into military service.

Throughout the land sirens sent people off the streets, scrambling for home at night. Blinds were closed and blackouts were imposed in case it wasn't a drill, and enemy planes seeking targets could spot and bomb our towns. Schools held air raid drills with teachers and students hiding under their desks.

All automobile production halted as factories were turned into arms manufacturing facilities. No one escaped the war effort as communities held scrap drives to collect unwanted metal that could be melted down and used as weapons. Harsh rationing of everything from gasoline to soap to butter was imposed. Food and gas stamps were issued to stop hoarding, and buying tires required special permission.

Small children pulled wagons around to gather milkweed, and scrap metal for the war effort. Grade school students saved their pennies to buy stamps for war bonds to help the effort. People from every strata of society, and every age group pitched in. Farmers, factory workers, retailers, professionals, and business owners, worked together to win the war.

Women became factory workers, and doubled their work to keep things going at home. Hollywood made upbeat films, and many actors, and entertainers even enlisted. Bob Hope and others entertained troops in the US and overseas to keep up morale.

It was unpatriotic to complain about the shortages, and everyone pulled together. Our freedom depended upon our military victories. Small flags with silver stars attached were posted in homes when loved ones entered the military, only to be replaced with gold stars if they were killed in action.

The war seemed as if it would never end for Leguire. It dragged on year after year as he moved through elementary and junior high school. It finally ended in 1945 as he finished the eighth grade. Final victory did not arrive, however, until after the United States had dropped atomic bombs over two Japanese cities.

But life went on. Working on the farm was hard work with lots of chores before and after school. There were cows to milk by hand twice a day, and animals to feed. It was made worse by Leguire's younger stepbrothers who were good at giving him most of the heavy work to do. He began to plan his next move.

Chapter Three

LEAVING HOME — THE TALLADEGA TRIP

LEGUIRE WORKED HARD, TAKING ON THE HEAVIEST PARTS OF the farming. It didn't take long for this ambitious young man to start developing plans to leave the farming to others. As resentment built up against his stepbrothers, Leguire decided to move on to a different life where he could have more control over his own destiny.

In 1946, when he was only 15 years old, he quit school during the ninth grade. Leguire hitchhiked 200 miles to the city of Toledo in northwestern Ohio near the Michigan border. Later he returned to Piketon, Ohio, where he landed a job loading railroad cross ties into box cars.

At age 17, Leguire decided to join the US Marine Corps. He made his way back to Franklin Furnace where his mother reluctantly signed his enlistment papers. Leaving from there, he traveled west to Cincinnati, Ohio where he enlisted on January 26, 1948.

The Marine Recruiter put Leguire on a train to South Carolina for recruit training. He graduated at Parris Island, South Carolina April 13, 1948, as a Private First Class. Leguire still remembers his Drill Instructors in Platoon 20, Second Recruit Battalion. They were Sgt. J. T. Kesler, and Cpl. D. R. Coffin.

From Parris Island he was transferred to E Company, 2nd Battalion, 2nd Marine Division. In October of 1948, his outfit left Camp Lejeune and traveled to Newfoundland, Canada. They returned in late November.

Shortly thereafter, in January 1949, His outfit began a Mediterranean Cruise through Europe which lasted five months. They traveled aboard the aircraft carrier, Philippine Sea and were transferred later to the destroyer, USS Vogelgesang. They returned to the United States in May of 1949 aboard a sea going tanker.

[Leguire now picks up his story in his own words....]

"My Talladega Trip"

One of my favorite memories before the Korean War was a trip to Alabama in 1950. I owned a big 1941 Buick two door sedan and on May 30, 1950, I drove my Marine buddy, Lonnie R. Brown, to his home in Talladega, Alabama, for the weekend.

The next day, my date and I drove to a lake outside of the city. We were returning on a dirt road, and I was driving too fast because I was aggravated with her. I was speeding around a curve and slid into a ditch.

The car overturned, and rolled over several times before it came to rest upside down. She was screaming, "My leg is broken!" As I crawled out the window, her feet and legs were turned toward the same window. I could see she wasn't hurt, so I took her by the feet and began pulling her out through the open window.

When I pulled her part of the way out her clothes were pulled up over her head. Quite naturally, I stopped pulling to take a look at her. She began kicking and yelling at me, so I pulled her on out. I started to laugh hard at her, and she got really mad. I had to hold her tight so she couldn't kick me.

My car was lying upside down, and gas was pouring out. Then I got a little angry because I had just filled the gas tank; I believe gas was about 16 cents a gallon back then.

After we calmed down, we looked around and saw a house several yards away up a steep hill. We slowly climbed our way toward the house and it was well after dark before I knocked on the door. When the door finally swung open, I was looking down the barrel of a 12 gauge shotgun! A colored man behind the gun said, "What do you want?"

He scared the 'crap' out of both of us that night!!

I told him we had overturned our car down the mountain a little ways from him, and we needed a ride into town to Lonnie's house. He laughed hard at us and said, "I thought you were revenuers." I told him I needed a big drink to settle my nerves after he pointed that shotgun in my face! He laughed again and drove us to town, but wouldn't let me pay him for it.

Leguire and his Buick

My car was totaled, but to this day I still have my Ohio tag for getting on base at Camp Lejeune. I had to catch a bus back to camp and was a little late, but didn't get into trouble after I explained what had happened.

To this day I laugh when I look at the pictures of my wrecked car, but I never saw the girl again. The Korean War broke out soon after, in June of 1950. Lonnie and I and some other Marines were shipped by train to San Diego, CA, and then to Japan before landing at Inchon, Korea.

I saw Lonnie Richard Brown once after Korea when he came by to see me and my family. I received the news of his death too late to be there for him, and heard later that he died of Leukemia. I have never gone back to visit Talladega, Alabama.

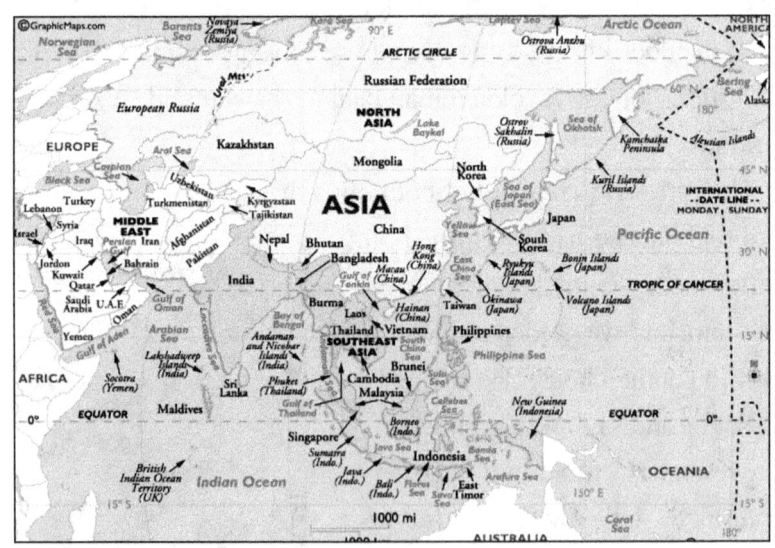

Asia

Chapter Four

PLANTING OLD GLORY — UNITED PRESS ARTICLE

MY OUTFIT STAYED IN CAMP LEJEUNE, NC UNTIL WE SHIPPED out in August of 1950. We loaded onto trains heading for California, and arrived at Camp Pendleton about three days later. We then loaded all our equipment aboard ships in San Diego and headed for Kobe, Japan.

In Kobe, we transferred all our equipment, supplies, and combat gear aboard another ship heading for Inchon, South Korea. It took our ship about three days to arrive at Inchon for the invasion.

I landed in Korea on Blue Beach, September 15, 1950 under Colonel Chesty Puller, and moved quickly to its Capitol, Seoul. As we neared Seoul, I spoke with Marguerite Higgins, a famous war correspondent working for the *New York Herald Tribune*. Miss Higgins was the only female war correspondent in Korea at that time, and

Marguerite Higgins

perhaps the best-known female war correspondent of all time.

On our way we saw a couple of ducks running loose. After catching them, my buddy tied them to his back pack. Later, we saw a Korean woman, and asked her to cook them for us. So we stopped for the night and she cooked duck and rice for us.

Oddly enough, that woman understood everything we said. When she heard me wish for it out loud, she heated a big pot of hot water, and I enjoyed the first bath I'd had since landing. That was around September 20th while we were in Seoul.

A few days later, as we moved through Seoul, I had the privilege of raising the Stars and Stripes over the roof of the American Ambassador's residence on September 26, 1950.

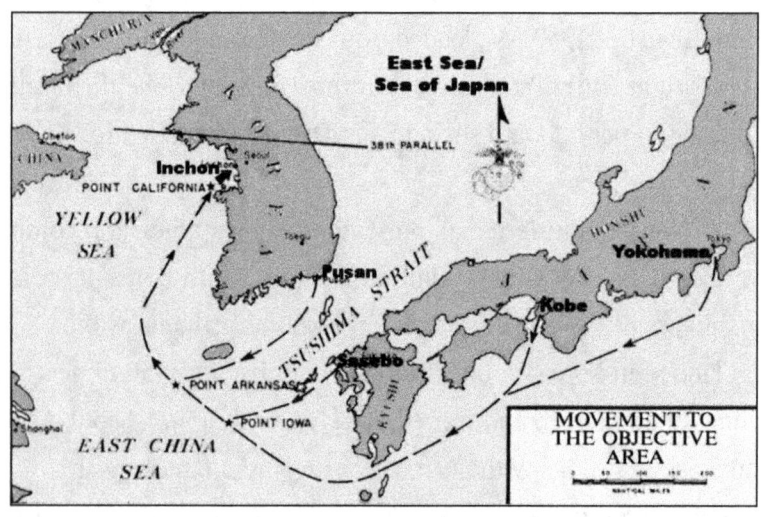

Movement to the Objective Area

United Press Article

The planting of the American flag in Seoul, Korea on September 26, 1950 was witnessed by Robert Vermillion, United Press Staff Correspondent and photographer Sgt. John Babyak, USMC. Their description of this isolated event became the official war record of the first American flag raised in the capital city of Seoul as follows:

> *Teletype: Seoul, Korea, Sept. 26* — American Marines hoisted the Stars and Stripes over the residence of United States Ambassador John J. Muccio in western Seoul today.
>
> A dirty-faced Private First Class, Luther R. Leguire of Dayton, Ohio, planted the flag on the porch roof of the building at 3:37 P.M.
>
> His act symbolized the capture of the Korean capital by United Nations forces only 11 days after they landed on the beaches at Inchon.
>
> The flag-raising ceremony was a gala and informal affair, without the usual solemnity that historians prefer. Marines first snipped wires on the flagpole in front of the residence to lower the North Korean Communist banner flying at its top.
>
> Marines lounged in the grass outside Muccio's residence, which had been looted by the Communists, as Leguire attached a thin flag pole to the corner of the porch roof.
>
> As the Marines watched the flag-raising at the Ambassador's residence, an enemy sniper fired one shot from the roof of Blazing Duk Soo Palace which housed United Nations offices before the Communist invasion the prior June. As it whistled over the Marines' heads, they said they were all for, "Getting that S.O.B."
>
> But their company commander ordered only one Marine to shoot, since there was only one sniper. The Marine's aim was true, and the sniper tumbled off the roof.
>
> Meanwhile, street fighting in the burning capital diminished to occasional flurries, and Marines moved through the city as rapidly as they could walk.
>
> The occupation of the entire city except for the northeast sector was assured when a new Marine regiment moved in from the northwest to link with forces already advancing eastward through the streets.

20   THE FORGOTTEN — Volume 1

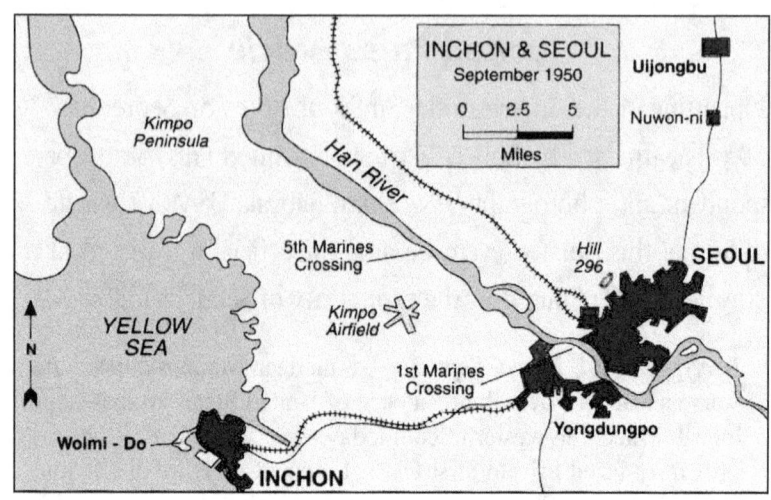

Movement to the Objective Area

Below are two photographs that Cpl. Luther R. Leguire keeps in his personal file. They were snapped by a Marine photographer as he planted "Old Glory" on the porch roof of the American Ambassador's residence amid sniper fire. He also has in his possession the actual tattered and bullet riddled North Korean Banner removed from the flag pole on September 26, 1950.

PFC Luther R. Leguire planting "Old Glory" on the porch roof of the American Ambassador's residence in Seoul, Korea.
Photo credit — Hqtrs. No. A3386 — Sgt. John Babyak, USMC

Chapter Five

ALL HELL BROKE LOOSE — I REMEMBER PRAYING

BY MID-OCTOBER, 1950, THE KOREAN WAR APPEARED TO BE nearly over. Most of North Korea had been captured by the American-led UN forces. By the end of October, the North Korean Army was rapidly disintegrating, and the UN forces had captured 135,000 prisoners. President Truman became anxious and rightly so. He ordered General MacArthur to be very cautious when approaching the Chinese border.

As they neared the Yalu River, **all hell broke loose**.

Both China and North Korea considered the United States to be their major enemy. The UN Command, under General Douglas MacArthur, was slow to appreciate the danger. MacArthur ordered his ground units to continue their offensive to the Yalu River (the border with China), and to cut the Chinese supply route.

On October 25, 1950, the People's Republic of China entered the war and huge numbers of Chinese soldiers poured across the border into North Korea toward the UN forces. Their heavy assaults halted MacArthur's drive.

[Leguire continues his story in his own words....]

After leaving Seoul on the west coast, our outfit headed toward North Korea. Quite suddenly, we returned to Inchon, loaded all the supplies aboard ships and headed for the east coast of North Korea near the Chosin Reservoir. We landed in Wonsan on November 7, 1950.

My outfit was ambushed that day during one of many battles surrounding the Battle of Chosin Reservoir, a major battle in the Korean War. The UN troops participating in those battles were nicknamed the "Frozen Chosin" or "The Chosin Few."

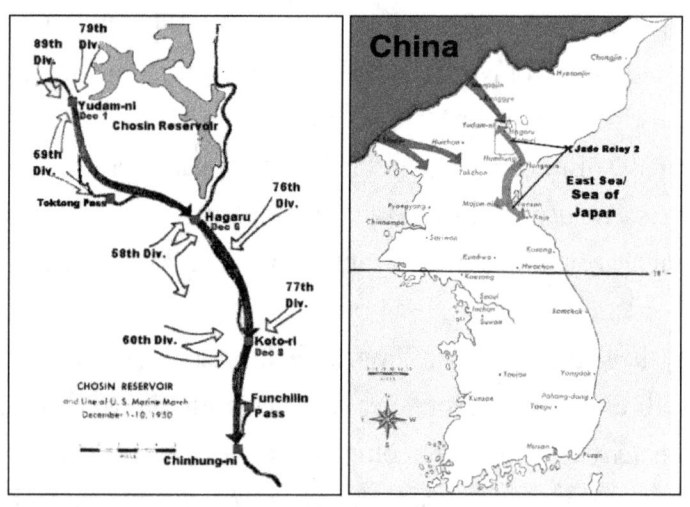

Chosin Reservoir fighting areas

My unit was on its way north from Koto-ri to help Fox company who were surrounded by the North Korean and Chinese forces. I was with E Company, 2nd Battalion, 1st Marine Regiment, under the command of Col. Chesty Puller.

It was fall, and the weather was turning colder. We were riding in trucks, moving up a steep hill on a road almost too narrow to travel. There was one tank in the front escorting several troop trucks.

On our way we saw many of our dead soldiers. Trucks had been blown off of the hill, and were scattered down the hill side. When passing this scene, I felt bad about all the destruction, and yet I wasn't scared; just in awe at what I was seeing.

The tank found an open place to turn into and get off the road. Our trucks could barely maneuver around it, but after passing the tank our convoy moved on up the hill. The tank backed out, turned and went back down the hill.

After the tank turned around and left us, we proceeded about a mile before the ambush began. I heard the familiar sound of the burp gun, and several of us jumped to the ground and returned the fire.

At first we couldn't see the enemy because they were hidden. We were out on the open road, and only a few feet from going over the hillside. Suddenly one of my buddies opened fire with the 50 caliber machine gun. He must have killed a hundred men before the enemy concentrated their fire on him, and he was shot between the eyes. He fell away from the 50 caliber and my other buddy took his place.

I don't know how long he stayed, but there were a lot of the enemy being killed as well as a lot of us. I looked away from him, and saw that the gooks were very close to us. We were on a mountain and they caught us out in the open. After a few minutes of trying to take some of them out with the machine gun, my second buddy was hit. Then another Marine took the gun and began firing. Bits of sand would fly up just by our heads. We couldn't go anywhere.

We weren't all in one group. I couldn't see the others because they charged up the hill fighting the enemy. We realized we had no place to go but up the hill into them. We captured seven of the enemy, shot them, and threw them over the cliff.

I fired at the enemy soldiers I could see. Then I was hit in the right leg by another enemy that I couldn't see. About five or six of the Marines gathered together with me. We were in an area about the size of a small house. They were still shooting right by my head. I remember the guys with me as they moved away from the truck.

One by one they all dropped and I was left alone, but alive. The last Marine fell across my legs. I had to push him off with my left leg so I could move. That was hard for me. We had not been together but for a short while. But when you are in battle you get real close real fast. We were like brothers as well as Marines!

They didn't all die right away. I can't remember what we said to each other, and I don't want to. A corpsman finally got to us. He gave me and others that were dying, several shots of morphine. I'm not sure how many. He did his best and moved on. I saw him no more.

All of my buddies that were with me were killed, and I was wounded. On behalf of all those that died during the ambush, I will say they died as bravely as they lived. They weren't crying or cursing; **they just went to sleep.**

After the others passed away, I dragged myself under a 6x6 truck that had been knocked out during the battle. Just as I started moving some North Koreans or Chinese shot at me twice right by my head. They were close but couldn't hit me. They kept trying to shoot me after I got under the truck, and their bullets and the sand bounced around my head.

I REMEMBER PRAYING

I can't remember feeling any pain, and I can't remember feeling sick, but I remember praying. It was under the truck that I began to pray to God and said, **"If You'll let me live and get home I'll serve You the rest of my life."**

I removed my rifle sling and made a tourniquet that I tied around my thigh. Then I ate my C Rations. I laid there for six hours before some friends found me, and administered some medical treatment. I would loosen the tourniquet on my leg now and then hoping to prevent gangrene from happening. I remember to this day the time I was hit. It was 11:45 AM. Why I looked at my watch at such an hour I will never understand.

I remember seeing Sgt. Jack Fichter when the medics arrived. We said goodbye to each other as they loaded me onto a truck with a stretcher. I was picked up around 6 PM and suffered a rough 25 mile ride on a 2½-ton, 6 by 6, Marine truck down the mountain to the Medical Station.

At the station I was treated with more medicine at about 8 PM and stabilized. This was nearly nine hours after getting shot. The next morning I was taken to the hospital ship, USS Consolation, where I was operated on, and placed in a leg cast. Then I was shipped to Yokasuka, Japan and put in a body cast that stayed on for several weeks.

They finally loaded us on stretchers for our trip home and placed us aboard a plane heading for Honolulu, Hawaii. While aboard the flight, I looked out of the window and saw the outboard engine catch fire. The pilot shut it down, and landed the plane on Midway Island. We waited there until parts were flown from Hawaii to repair the engine.

"I was in the Mare Island Hospital on Midway for my 20th birthday December 18, 1950." We eventually arrived safely in Hawaii, and our injuries were treated at the Tripler Army Base General Hospital for a few days, before moving on to California.

In California, four of us were placed aboard a small twin engine aircraft. As we lifted off one of the engines failed, and we turned and landed immediately. This, believe it or not, happened three times on three different aircraft! I was so scared after the third failure that I asked for a taxi ride across the United States to Jacksonville, Florida. This was good for a laugh, but not an option. Finally, the fourth plane got us safely to Corpus Christi, Texas.

Once there, we were placed on a larger plane bound for the Naval Air Station (NAS) in Jacksonville, Florida, and we arrived sometime in January 1951. I stayed in the hospital at the NAS until I was transferred to the Marine Barracks at the Charleston, SC naval base.

Let me respectfully pause to say that I salute all of the Naval Doctors and Corpsmen who so graciously tended to me during the time from the battlefield to the day I retired. They had to clean and bathe me until I could handle it myself. They are the best I've ever known in my 78 years.

While at the Charleston, SC naval base, I found out the Navy Department would let me retire 30 days early if I had a job either waiting for me or promised on the outside.

The Florida Highway Patrol had a job waiting for me as a radio operator. So I decided to ask the Colonel at the Marine Barracks to let me out 30 days early.

I will never forget what he said to me as I stood in front of him, supported by my cane. He said, "Marine, you stick around and earn your money! You are dismissed!"

I replied "Yes, Sir!" but I was not happy about it. I still laugh every time I mention this to anyone. As it turned out, I didn't get the Florida Highway Patrol job and there were no other jobs available.

Shortly before retiring I was told I had been promoted to Corporal grade on April 18, 1951. On November 30, 1951 they retired me out of the Marines.

My Marine discharge shows my total separation pay was $26.53 plus $14.04 travel pay to Jacksonville, Florida. I received a total of $40.57 at the end of my active service time of three years, ten months, and four days. Later, in 1952, I was given another $300.00 Mustering Out pay through the Military Act of 1952.

I was married ten months before my discharge on January 24, 1951. We settled down in Jacksonville, Florida, but jobs were scarce. I couldn't find a job because I was 50 per cent disabled. Several years later a law was passed that stopped employers from discriminating against the disabled.

I went to see my friend, Congressman Charlie Bennett. He called the US Post Master, and I was given work that lasted until January 1, 1953. After that job I sold Kaiser Frasier automobiles. I sold only one car, and made $100.00 which was equal to about two weeks salary at the most.

A short time later, I went to work for a Navy Chief Petty Officer who owned a propane gas business. Despite wearing a leg brace, I filled and delivered bottles of gas weighing from 220 to 325 lbs. to trailer parks. The Chief Petty Officer sold his business to me after I had worked there a few months. I continued in the propane business for nearly two years before going broke, and losing it all.

Once again, in June of 1953, I went to see my dear friend Congressman Charlie Bennett. He put his name on my Form 51, and I was given a job in procurement at the Naval Air Station (NAS) Cecil Field, Jacksonville, Florida.

I worked there nine years before resigning my job at NAS to start a new church in Lake City, Florida. My leg slowly recovered, but I stayed in a leg brace for about six years.

Chapter Six

THE FAITH JOURNEY — GOD MOVED ON ME

AFTER I RESIGNED MY JOB AT THE NAS, I STARTED PAINTING houses. I built my own home in Jacksonville in 1959. The connection between my combat injury and near death experience in Korea, my battlefield commitment to God, and getting into feeling the Holy Spirit and preaching led to studying God's word.

I attended church for 11 years, and became a minister by studying and researching the Hebrew translation of the King James Version of the Bible. I wanted to know for myself what the Bible taught. I didn't go to seminary. I call them cemeteries anyway. But that's just me.

In January 1962, God led me to start a new church in Lake City, Florida. I have been its Pastor now for over 46 years and I thank God for His goodness to me through the years

As Pastor I worked with my hands like Apostle Paul in the New Testament. I wanted to experience God in my life as I learned His word. I wanted to know what I was talking about and why.

I could believe God was real because He saved my life in Korea. I remember as a young man aboard ship while in the Marines, I would stand on the back deck, look up to the skies and say, "I know there's a God." In my own way I felt God's presence, and I would weep.

I started to believe as a kid when times were tough. The problem was I dearly loved God but hated the church. I didn't really know God but I got baptized anyway. They cut a hole in the ice on the Scioto River in southern Ohio, and put me into it. When they brought me up, I wasn't even cold.

I think that's the reason the Lord had mercy and saved me in Korea. I didn't know how to be a Christian or anything like that. I'm telling you the story of what happened to me in Korea, and how God called me. I don't have anything to be ashamed of or to apologize for.

One day an amazing thing happened as I was walking down the hall in a hospital in Jacksonville. I was on my way to visit some people when suddenly a mother stopped me. She knew me but I didn't know her. She said "My son is dying, would you pray for him?"

The nurses in that hospital said he was dead. He wasn't moving, breathing, or anything. He'd been in a car wreck and crushed his head. I was scared to death.

I went into his room, touched the boy's feet, and began to pray, "If it's your will God, you heal him." That boy sat straight up in bed. His mother rushed in and said to him, "Do you know who I am?" He said, "Yes, you're my mother."

Now that did something for me.

You know I don't go around saying I raised the dead. I don't do that. But I know when I did what God wanted me to do, God healed that boy instantly. He was in the hospital and given up for dead. They all said he was dead.

The nurses were leaving, and his mother was out in the hall when the father arrived. He said, "Son, do you know who I am?" And the boy said, "Yes, you're my Dad." He could sit up and walk! Well, I didn't let that go to my head because I knew who healed him. God was preparing me.

I could go into a lot of things He did to prepare me. But God is real. He is not just someone you call on when you need Him. People say, "Oh yeah, I believe in Him." But God wants to know that you truly believe in Him. You have to prove it to Him.

Tests are the way God watches out for us. God discerns our intentions, attitudes and motives through tests the same way a guard in combat discerns the intentions and motives of those approaching the camp.

All through my life God has been testing me to keep me safely inside His spiritual walls. I call it walking by faith.

GOD MOVED ON ME

One special night God moved on me, and called me to go to Lake City. A light shined in my room at 2:30 in the morning. That was the most amazing thing I'd ever seen. That whole room filled with light, and I just fell on my face before God and began to pray.

All I could hear in my heart was, "Go to Lake City; Go to Lake City." I could tell you some stories that would be hard to believe, but I believe God. If you don't believe God, don't play with him. Lots of people say they believe, but don't.

When God laid it on my heart to go to Lake City, I had never been there before. Interstate 10 from Jacksonville to Lake City did not exist in those days. There was only one tried and true route and that was US 90.

So I took off one day, and I drove over to Lake City. It was between pay days so I first stopped at a Sunoco oil station and got $5.00 cash on my credit card. I went to Lake City with only $5.00. That's all the money I had.

While I was looking around, something kept saying to me, "Why don't you look at the old Presbyterian Church?" The structure was unused because the congregation had moved to a new building. During the Civil War it had been used as a hospital.

So I went to visit Dr. Montgomery, the Presbyterian minister. He had a deep English voice, and had studied at Oxford University. He was no fool. He was a Chaplain in the First World War, and had been in Lake City 40 years. He was the kindest man I ever knew.

When we met, I said, "God has called me to this city." I proceeded to tell him my story, and then asked if I could rent the old church building.

After my request I added, "Dr. Montgomery, will you pray that God will bless my ministry?" He did, and after his prayer, he said, "Brother Leguire, call me Monday. I'll meet with the church board and see what the rent will be."

When I called back he said, "If we pay for the lights, water, and insurance, can you pay five dollars a week?" I said, "I can," and thanked him, and the church board. I later found out Dr. Montgomery had told the board, "If Brother Leguire can't make a go of it in this city, no one can!"

They were really good to me by letting me rent their old building. That's really how I started our church in Lake City, Florida. I paid them five dollars a week for four and a half years, and then we built our own building in 1966.

When we first started our church here, I was making good money. In those days I was a white collar worker at the NAS Cecil Field in Jacksonville, earning $3,700 a year.

That was big money in the 1950's. I built a new home in 1959 that cost me $8,900. I still have the house plans and the check to prove it. I also bought a new car every year. So we had everything we needed.

I spent $40 a month driving from Lake City to the commissary at NAS Cecil Field. I only bought $40 worth of groceries each month. A gallon of milk cost 39 cents. I drove back and forth for four and a half years before resigning my job at NAS Cecil Field.

I remember the time after I had sold my new car, and bought an older model. It was a quart low on oil. I could buy oil on the naval base for only 15 cents a quart that cost 25 cents in Lake City. The car also needed a grease job and oil change but I couldn't afford it.

I was driving down Normandy Blvd. on my way to the commissary at NAS Cecil Field, when WPDQ Car 60 came along behind me. That car was famous in Jacksonville, so I turned on my radio to see if it worked. The first thing I heard was, "WPDQ Car 60, where are you?"

I heard the driver say, "I'm following this 1961 Chevy Station Wagon. If the driver pulls over to the side of the road he has just won a grease job and an oil change." I pulled off the road so fast the WPDQ car driver almost ran into me.

He said, "If you've got this certain number on your driver's license, you've won $500." I didn't win the $500 but I got the grease job and oil change free. I also had my $40 in groceries and God had again showed me what He could do.

No one knew we needed these things but us. God has always provided for us. I drove back to Lake City thanking God for supplying my need. Now what are the chances of that happening? I had just turned the radio on in my car, and here came the blessing. Another thing I have never worried about is money. I don't have a retirement plan because I don't plan on retiring. I never give that a second thought.

We bought a house and moved to Lake City in June of 1963. One day a preacher and a friend came up from Miami. His name was Henry Dunn. They stopped while on their way to Louisiana. He said, "Brother Leguire, how are things going?" and I said, "It couldn't be any better." Then I invited him and his friend to spend the night with us.

The next morning after breakfast, my wife, Margaret, told me that we didn't have any more food in the refrigerator. In fact there was no more food anywhere. We had just eaten our last bit of food. That's when I said, "Don't say anything to anybody because the Lord knows about it."

I waited until after they had gone, and went over to the old Presbyterian Church building. It was cold because there was no heat in it.

I knelt down in prayer but instead of saying, "God, I need help," I was thanking Him for allowing me to be there. I was thanking Him for giving me a chance to serve Him.

While praying, rap, rap, came a knock on the door. I got up, went to the door, and my wife said, "Look at what I found under Brother Dunn's plate after he left." She showed me two 20 dollar bills he had hidden! Well, I guess you know we both began to weep.

I kept working in Jacksonville for one and a half years. I was a self employed house painter and also drove a school bus for about seven years. I didn't want to live off the church, even though we had many members, but the Lord changed that.

Here's how He did it.

I had been painting a house for a contractor, but because of rain I couldn't get a draw. So I was out of money. One day while painting on a scaffolding three stories high, I became so physically sick I had to lie down. It wasn't an actual voice, but someone called to me and said, "How will you get down from this scaffolding without me?"

God's silent voice was telling me that He would supply the needs of the church. I said, "Lord, I'll do whatever you want if I get down off this scaffolding." So I closed the job out and did no more work like that.

Here I had been holding up God's work and didn't know it. I was doing the work of the Lord, so it didn't bother me. I knew God would take care of his people. But you know how a man thinks he knows what is best. I thought I'd quit after the church grew to a certain size, but after that I started accepting a salary from the church.

I drove a school bus in Lake City during the 1960's and early 70's. Several public office holders like the Sheriff, State Attorney, and police officers rode on it when they were young. In those days the drivers had paddles on their buses, and I have paddled some of those officials.

Just for a note of humor; one day I was speeding toward my office at the church. A police officer followed me into my parking space with lights flashing and sirens blaring! After greeting me, he said "You spanked my bottom, and now it's payback time." We both hugged each other and laughed, and there was no ticket.

I not only drove a school bus, but I visited hospitals twice a day. I was changing clothes three to four times a day, but I loved every bit of it. I didn't mind getting out there in the hot sun scrapping a house or driving a bus for the Lord.

The church was just beginning to grow. Our number went up to 65 the first year and then back down to five. The folks just moved away. I thought I'd failed God somewhere.

There was an old man who just visited the church every day. One day when he stopped by I told him, "I've got to go back, and sell my house in Jacksonville. I can't stay there anymore."

I had tried to sell my house the year before. So I just went home, took the sign down, and said, "Well Lord, if you want me in Lake City, you'll have to sell my house here first."

And that very day I got a call from a guy that had looked at it the year before. He was in New Jersey in the Navy. He said, "Do you still have your house for sale?" I said, "Yes sir!" He said, "I want to buy it." I was gone for three months and then I returned to start the church back up again.

I started holding services again and the numbers began to grow. Slowly we made our presence known throughout the community. I was working, driving a bus, visiting the sick, teaching scripture, and holding worship services. It wasn't easy and we had our difficulties, but we hung on.

After I'd been here four years, nearly all knew me and my beliefs. I had quit working. One day I had $300 with me to buy a new clothes dryer for our home. There was a nice piece of land owned by a man here in town. He wouldn't sell it to anybody, and I decided to check it out.

I stopped and said, "Mr. Summers, I want to buy this piece of property. Will you sell it?" He said, "Lots of people want to buy it, and I've never even had it up for sale, but Preacher, I'll sell it to you." That's how I bought the property we're on now. I paid $7,500 for seven acres in 1966. Do you have any idea what this property is worth now?

I put my $300 down for the property, and said, "I guess there won't be a need for a new dryer today." Then I said, "I've got to go look for some clothes line." A little later God provided a way to get the new dryer too.

Everybody asked me how I was able to buy this land. There were a lot of builders who had been trying to buy it. It's right on the corner, and there are big beautiful trees on it. It was another of God's gifts to us.

We built our first church in 1966. It was a prefabricated building from Illinois, and I had to borrow the money. A friend of mine worked in the First National Bank. He knew I needed the money, and suggested I talk to an attorney who was one of the most powerful men in this city.

So I went to see him, and asked for a loan. When I told him I didn't have any collateral he said, "I'll tell you some-thing preacher. I have always been leery of anything that started with "P" like Preachers, and Painters, and Plumbers." So I said, "How about Politicians?"

He laughed real loud and then wrote me out a check for $12,000. We got along fine. He'd buy me or my family lunch right now, if he saw us in a restaurant. We paid off the loan and never had to borrow money again. We later built a new sanctuary next to the older one.

I received building plans for a prefab Church in the mail the next week. I had already ordered it for $8,000. It was our first church building, and we decided to put it up ourselves.

I poured the foundation and had everything ready. The manufacturer said, "We'll have it there for you next week." Next week came and went but no building. So I called them up a little after midnight.

I said, "Tell you what I'm going do; I'm going to give you until 8 o'clock in the morning, and if you're here at 8:05 you're taking that building all the way back to Illinois with you because I'm not buying it. I listened to you telling me what you're going to do. Now I'm telling you what I'm going to do."

They drove all night. One man sat on a seat, and the other man sat on a five gallon bucket. They switched to give each other a break. They hauled that thing all night. At five minutes before 8 they were having a hard time finding us so they called me. I said, "You only have 5 minutes." They said, "Well now Brother Leguire we're here, we just can't find where you are." I said, "Okay," and gave them directions to our place.

The mayor of Lake City at that time was a friend of mine, and allowed me use the City's equipment. There was nothing here but dirt. They put sidewalks down, and then ran a gas line all the way out to the church. Not my church; God's church.

There were no houses or big buildings between the church and the last street in town, so he sent his crew out here to help me clear my property. That was when the Mayor was boss. He was boss and he helped me

The church grew and doubled in size. Then in 1972, we built a brand new building next to it. The walls are strong. They are 13" thick concrete block with steel reinforcing bars running down through them. Every so many feet I placed steel bars, and then poured concrete down inside the blocks as they handed them up to me.

These walls would be mighty hard to knock down. The pilasters outside the church stabilize it. God could take it down, but that's about all because I put steel rods into it.

We held fundraisers. In just nine months time we sold $19,000 worth of peanut brittle to pay for our present building. After we finished the church building, people would still buy peanut brittle if we'd sell it. I said no, we won't sell any more. We built the church and it's paid for.

Whenever we sold meals to raise money, we didn't do like the others, and just pile food on a dinner plate. No, we had sectioned plates. We gave them a real dinner, and then carried it to them.

Then the head of Armco, Standard Oil at the time, came and offered money if we'd pass out glasses to the houses around the church. Look how God blessed us. We passed out glasses to advertise for his Oil Company and got paid.

I also had a contract with a book publishing company. We bought Bibles for $3 apiece that retailed for $20. Then we gave a Bible to everyone who gave a $20 donation to the church. We kept $17 and they received a $20 Bible. Those are some of the things we did.

There are many more stories about our Church that you wouldn't believe. I'm just telling you a few.

Chapter Seven

YOUTH BOOT CAMP — LEGUIRE'S LEGACY

AS CHURCH PASTOR, I'VE TAKEN OUR YOUNG BOYS AND GIRLS to Camp Montgomery for 43 years. Their ages run from 11 to 16 years old with many over 18. I call it U.S. Marine Corps youth training. Camp Montgomery is located west of Lake City, and sits along the banks of the Suwannee River

During camp they are taught basic Marine Corp Ethics including how to fall in and stand at attention, calisthenics, running, and getting up two or three times a night. They are taught to eat everything put on their plates, and to eat all of it! If I find out they don't like a certain food...that is what they are served.

They are taught to swim, some by being thrown into the springs and river. All have learned how to swim and dive, and to love the water. They learn to say, "Yes Sir" and "Yes Ma'am," and "can't" is not an acceptable reply. When they approach one of the Camp Counselors, they must stand at attention until "At Ease" is given.

Camp Montgomery is out in the country with all kinds of snakes, panthers, black bears, wild cats, and more. When a counselor says, "Freeze!" they know from their training that it signals danger or something else as serious.

There is a set of steps at the camp without a hand railing. They are told to put their hands on an imaginary hand rail along the steps, and to stand there with their hands on that rail until they are told to let go. When they are told to freeze they stop dead still.

While upstairs where I slept, I recall one night at 2:30 AM watching two of the older boys slipping up to the kitchen to get some drinks that were sitting out on the porch. Just as they leaned forward about six inches from the bottles I said, "Freeze!" And they stayed in that position until I got to them.

All those kids tried to beat the system, and they still say I never slept, which is almost true. When the first campers grew up, married, and had children, they made sure their children were ready for camp. Since they knew the routine, I didn't have to train the second and third generations as much as I did their parents and grandparents.

The following confidential letter was sent to the author from a grateful survivor of Leguire's famous camp. It sheds light on why his youth training program was so successful.

Dear Mr. Cummins,

May I have the honor and opportunity to say a few words about Mr. Luther R. Leguire? I pray you will give me a few minutes of your time to tell you of this great man.

My name is Karen M. Oglesby-Bareno of Villanueva de la Canada, Madrid, Spain. I am the wife of Don Manuel Bareno Lopez, Retired Major pilot of the Spanish Air Force, and currently Captain of Iberia Airlines in Spain.

I have known Mr. Leguire since I was 11 years old (32 years to be exact). I have had the honor of growing up, and knowing this wonderful Marine who has been more to me than a Pastor, friend, and confidant. He has also been much like a father to me. I am filled with so much gratitude for having the honor of sharing practically my whole life with Mr. Leguire, and for having this man in my life and by my side. How fortunate I have been.

Mr. Leguire's unmistakable love for God, his country, family, and friends has been his whole and true meaning for living. He has always put others first, before himself.

I have an 18 year old daughter (Alejandra) and I have had the luck to have instilled in her the same upbringings and teachings that Mr. Leguire taught me; things such as integrity, morals, honesty, trueness, dependability, and so much more.

We are grateful for Mr. Leguire's Marine training, and the example of how he lives, and has lived his life in complete TRUENESS, I and so many others are who we are today because he succeeded in making us fine men and women.

I guess you can say we were raised as close as you can get to being a Marine. I can't resist giving you an account of how it happened. With much fondness, I want to recall for you some of those moments at, "YOUTH (BOOT) CAMP."

Every year (for one month out of the year) we used to have youth camp down on the Suwannee River in Branford, Florida. It never gets old telling this story because it will always be in my heart and I really miss those days.

Mr. Leguire's strategy was to make us into fine young men and women. To teach and give us responsibilities that made us into tougher — better young people.

It all started at 0600 hours every morning with the ringing of a very large metal bell that sat in the middle of camp. By 0630 we had to be dressed and groomed (oh yes! there was inspection to see if we fulfilled our tasks) and make it on time to line-up in the center of camp to start our morning exercises.

Those normally never lasted less than 30 to 40 minutes (of course at our ages it seemed two hours instead) unless you were late or didn't pass inspection. Punishment was a run to the gate and back which was only 1.5 miles (but of course this was pure torture to us because it felt like 10 miles instead).

Then after exercise it was breakfast hour. We always ate good down at camp but we were also taught that whatever was put on your plate — you ate. No wasting food and if you were caught cheating, guess what — you earned yourself double rations. And God forbid if you sassed, protested or talked back because it was another trip to the gate and back. We were taught quite well.

Then there was KP Duty at breakfast, lunch, and dinner. KP Duty proceeded like this: Hot Water — Cold Water — Hot Water — Wipe. Those dishes and kitchen pans had better shine and squeak of cleanliness. If fact, I remember Mr. Leguire would come in after we were done, and run his fingers down a plate. Let me tell you it had better squeak, and not have a drop of oil on those pans and pots either.

If not — well the process was repeated (and I mean everything had to be redone) again and again until we had it right. Mr. Leguire always believed in good hygiene and cleanliness) and you know to this day I am the same way.

Then there was cabin inspection, and let me tell you when you cleaned your cabin it better be like the kitchen. Our cots had to be made soooo tight that a coin would bounce off of it. Yes, we passed inspection everyday, if not — a trip to that horrible gate again.

We had sooooo much fun and despite what it sounds like, you know I never complained, I actually learned how to clean well, and I mean wellllllllllll.

At the end of the afternoon we were rewarded by floating down the river and going to the springs. Mr. Leguire always watched out for us to keep us out of harm's way. He always had our backs against danger, and in all those many years we never had bad accidents.

Big ole' teddy bear, he is; but don't tell him I said that. After dinner every night he would load us all up and take us to buy all the candies, ice-creams, and sodas we wanted. After sunset we would laze around and he would tell us stories and just be with us.

As you can see, Mr. Cummins, we were taught discipline, respect (Yes Sir—No Sir), hygiene, cleanliness, responsibility, and so much more. And yes, it really worked. It worked so well we looked forward to going back every year, so what does that tell you?

In other words, Mr. Leguire, is, has been, and always will be our mentor, rock, and foundation. He has never failed or ever let us down. He has watched my back many times over the past 32 years. My life has been touched by a True Hero.

I thank him a trillion times over for his vision, and for listening to God, and to his heart!!!! A True and Genuine Gentleman who has looked to our futures squarely and proudly!

Consider this: I was raised around military. My father was Navy and we shifted from base to base. Then I married an Air Force Major whose father was an Army General. I have seen and talked with many that have been in wars and suffered losses.

My point is this! Despite the horrible circumstances that Mr. Leguire suffered as he lay in that trench with a shattered leg in Korea, while watching his compatriots and friends shot down right before his eyes, he was a Marine first.

His desire was to continue serving his country, no matter if the price meant his life. He was there to do his job and do his duty, and not once was he ever ready to just give up and let go. You can see his determination and strong will to be a good soldier and to serve his country well, in the man he still is today. I salute Pastor Leguire with pure admiration.

> *Once again Mr. Cummins thank you for your time and I hope that this has helped you to see another part of a True Hero. If you have any questions, please feel free to get in contact with me anytime.*
>
> *Most sincerely,*
> *Karen Oglesby-Bareno*

LEGUIRE'S LEGACY

The most enduring image of Pastor Leguire is that of a strong leader with a church membership who dearly love him. The great depression, WWII, four years in the Marine Corps including seven combat weeks and a near death experience in Korea, and six years in a leg cast became precursors to a life dedicated to God.

Leguire's fondness for military style leadership is best expressed in Karen's Youth Boot Camp letter printed above. His Marine training based upon subordination, compassion, and order, have served him well during his life and ministry.

As a faithful disciple of Jesus he inspires by example. He is a humble man of God who walks daily among his church flock and community fulfilling his special calling. This has truly been an epic journey for a man who has traveled from sharecropper to battlefield to pulpit during his lifetime.

> "Leadership is lifting a person's vision to higher sights, the raising of a person's performance to a higher standard."
> —Peter Drucker

THE FORGOTTEN WAR

Our Heroes' Stories

TIME MAGAZINE MAN OF THE YEAR 1951
Name: American. Occupation: Fighting-man.

North Korea Invades South Korea

Sometimes war seems inevitable, creating many questions impossible to answer using love, logic, and common sense. Perhaps the following riddle holds the answer to South Korea's dilemma, as it succinctly defines the eternal conflict between good and evil....

> "How does a good man live in peace, if his evil minded neighbor doesn't want him to?"
>
> —*Author Unknown*

Chapter One

THE INCHON INVASION

BY 1950, AMERICA HAD SURVIVED A DEVASTATING WORLD WIDE depression followed by an overwhelming World War II. Although the people were steeled by the hardship of the depression, and the war, they still believed fervently in this country.

Americans put aside their differences in the Korean War and worked together to win it. Although the war was distasteful, the American people stuck with the President because it was their patriotic duty.

Every strata of society from young to old pitched in, just as they had during WWII. You never heard prominent people on the radio belittling the President or the war. Unlike today, no newspaper or media pundit would have dared complain about what we were doing to win the war.

Much was different back then because we went to war to win. Unfortunately, Communism had replaced Nazism, Fascism, and Socialism. Its leaders, with deadly intent, decided to strike America's resolve and way of life.

The North Korean government, with the support of China and Russia, attempted to unify the divided Korea into a single Communist country. When the North Korean Army attacked South Korea, the United States decided to take action.

Backed by a resolution from the newly created and fledgling United Nations, President Harry Truman mobilized US military forces still in Japan, and General MacArthur was named UN Commander-in-Chief for Korea,

On being told of the outbreak of large-scale hostilities in Korea, President Truman ordered MacArthur to transfer munitions to the Republic of Korea (ROK) Army, while using air cover to protect the evacuation of US citizens.

Despite US intervention, the victorious North Korean forces advanced southwards. By September, only the area around Pusan — about 10 percent of the Korean peninsula — was still in coalition hands.

With the aid of massive American supplies, naval and air support, as well as ground reinforcements, the UN forces managed to stabilize a line along the Nakdong River.

This desperate holding action became known in the United States as the Pusan Perimeter. In the face of fierce North Korean attacks, the allied defense held and the North Koreans failed to capture Pusan.

In order to alleviate pressure on the Pusan Perimeter, General MacArthur received permission from the UN for an amphibious landing far behind the North Korean lines at Inchon. The violent tides and strong enemy presence made this an extremely risky operation.

This daring amphibious operation was conceived by General MacArthur. Though strategically tempting, Inchon was a tactically challenging amphibious target, with long approaches through shallow channels, poor beaches, and a tidal range that restricted landing operations.

Marines fight through Seoul in 1950 offensive

Forces that gathered for the Inchon invasion included the First Marine Division, the Army's Seventh Infantry Division, a few ROK military units, dozens of Navy warships, and virtually every available amphibious ship. Most of the Marines had recently arrived from the US, while the rest had been withdrawn from the Pusan Perimeter defenses.

Preliminary naval gunfire and air bombardment began on September 13, 1950. The Inchon Invasion began when the 1st and 5th Marines went ashore on
the morning of the September 15th. Resistance and casualties were modest, and initial objectives were quickly secured.

After hurling itself fruitlessly against the Pusan Perimeter for nearly a month and a half, the weakened North Korean Army was suddenly confronted with a grave threat to its rear. US Marines had landed at the western port city of Inchon, near Seoul, and were poised to move inland to retake the capital and decisively cut the already tenuous North Korean supply lines.

Over the next several days, as supplies and troops poured ashore at Inchon, the Marines moved relentlessly toward the capital city of Seoul. Kimpo airfield was taken on September 17th and supported operations two days later.

On September 16th, the Pusan Perimeter's defenders, who were located a hundred miles southeast, went on the offensive. After resisting for a few days, the isolated North Korean Army retreated and progressively collapsed during the rest of the month.

On September 27th, the US Army units that were moving southwards from Seoul met those coming up from Pusan. On September 29th, after days of hard street fighting, the capitol city of Seoul was returned to the South Korean government.

Korea was a brutally personal war as the Veterans learned the hard way. Their patrol actions were vicious and fighting on the ridges was a nightmare with hand-to-hand combat. Both cooks and clerks knew their M-1's as well as their field kitchens and typewriters. We must never forget what our heroes suffered through during that 'forgotten war.'

Chapter Two

CHINESE INTERVENTION

THE UNITED NATIONS TROOPS DROVE THE NORTH KOREAN troops north past the 38th parallel. The US then made amphibious landings at Wonsan and Iwon on the East coast, which had already been captured by South Korean forces advancing by land. The UN forces crossed into North Korea in early October, 1950.

The US led offensive greatly concerned the Chinese, who worried that the UN forces would not stop at the Yalu River, the border between North Korea and China. President Truman ordered General MacArthur to be very cautious when approaching the Chinese border.

The Eighth US Army, along with the South Koreans, drove up the western side of Korea and captured Pyongyang on October 19th. By the end of October the North Korean Army was rapidly disintegrating and the UN took 135,000 prisoners.

On October 8, 1950, the day after the American troops crossed the 38th Parallel, Chinese Chairman Mao Zedong issued the order to assemble the People's Volunteer Army (PVA). Seventy percent of the members of the PVA were Chinese regulars from the Chinese People's Liberation Army.

Chairman Mao ordered the Army to move to the Yalu River, The Chinese made contact with American troops on November 1, 1950. Then all hell broke loose when the Chinese seemed to come out of nowhere, as they swarmed around the flanks and over the defensive positions of the surprised United Nations troops. Thousands of Chinese soldiers attacked from the North, Northwest, and West against scattered US and South Korean units that had moved deep into North Korea.

Mao also sought Soviet help though direct assistance was mostly limited to providing air support. The Soviet MIG-15's in PRC colors posed a serious challenge to UN pilots. In one area nicknamed "MIG Alley" they held local air superiority against the American-made F-80 Shooting Stars until the newer F-86 Sabres were deployed.

In late November, the Chinese struck in the west, along the Chongchon River. They completely overran several South Korean divisions, successfully landing a heavy blow to the flank of the remaining UN forces. The US 8th Army escaped complete annihilation by the Chinese, mostly due to the successful but very costly rear-guard action of the Turkish Brigade at Kunuri which slowed the Chinese onslaught by three to four days.

In the East, at the Battle of Chosin Reservoir, a 30,000 man unit from the US 7th Infantry Division and US Marine Corps was also unprepared for the Chinese tactics and was soon surrounded. They eventually managed to escape the encirclement, albeit with over

15,000 casualties, and after inflicting heavy casualties on six Chinese divisions.

The US forces in Northeast Korea, who had rushed Northward with great speed only a few months earlier, were forced to race Southwards with even greater speed, and form a defensive perimeter around the port city of Hungnam.

On December 16, 1950, President Truman proclaimed a national state of emergency in order to fight "Communist Imperialism." The entrance of China into the war revealed the truth about the war. It was not about uniting North and South Korea. It was about choosing between worldwide Communism Imperialism and Freedom.

Facing complete defeat and surrender in late December 1950, a major evacuation was carried out. One hundred ninety three shiploads of American men and material were evacuated from Hungnam Harbor in Eastern North Korea. About 105,000 soldiers, 98,000 civilians, 17,500 vehicles, and 350,000 tons of supplies were shipped to Pusan, South Korea, in orderly fashion.

As they left, the American forces blew up large portions of the city to deny its use to the communists, depriving many North Korean civilians of shelter during the winter.

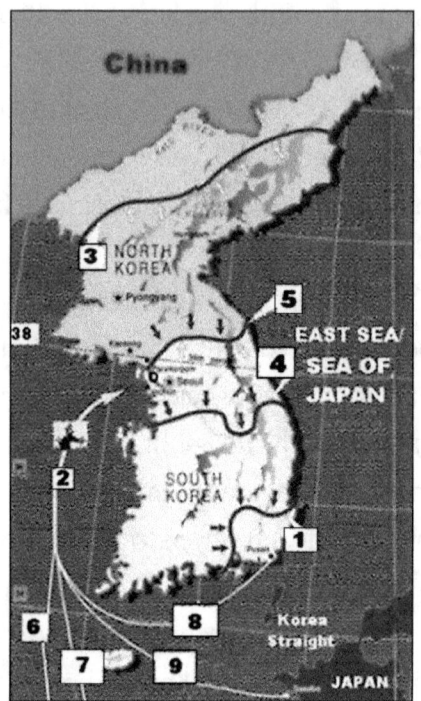

Date and Activities Chart of Korean War

1 **Pusan Perimeter** (August 5-September, 1950)
2 **Inchon Landing** (September 15, 1950)
3 **Northern Line** (November 24, 1950)
4 **Southernmost Advance** (January 24, 1951)
5 **Approximate Line during Armistice Talks** (June 15, 1951)
6 First Marine Division from Kobe, Japan
7 Seventh Army Division from Yokohama, Japan
8 Fifth Marine Regiment from Pusan, Korea
9 MacArthur's Command Ship, Mt. McKinley from Sasebo, Japan

Chapter Three

STALEMATE AT THE DMZ

DURING JANUARY OF 1951, THE CHINESE AND NORTH KOREAN forces struck again in their 3rd Phase Offensive known as the Chinese Winter Offensive. The UN resistance crumbled and their troops retreated rapidly to the south. Seoul was abandoned and was captured by communist forces on January 4, 1951.

To add to the Eighth Army's difficulties, Lieutenant General Walton H. Walker was killed in a Jeep accident. He was replaced by a World War II airborne veteran, Lieutenant General Matthew B. Ridgway, who took immediate steps to raise the morale and fighting spirit of the battered Eighth Army, which had fallen to low levels during its full retreat South across the 38th parallel.

Nevertheless, the situation was so grim that MacArthur mentioned the use of atomic weapons against China, much to the alarm of America's allies. UN forces continued to retreat until they had reached a line south of Suwon in the West, Wonju in the center, and north of Samchok in the east, where the front stabilized.

The Chinese People's Volunteer Army (PVA) had outrun its supply line and could not go beyond Seoul because they were at the end of their logistical supply line. All food and ammunition had to be carried at night on foot or bicycle from the Yalu River on the Chinese border.

In late January, finding the PVA lines in front of his forces deserted, Lt. Gen. Ridgway ordered reconnaissance in force, which developed into a full-scale offensive called Operation Roundup. The operation proceeded gradually and made full use of the UN force's superiority in firepower on the ground and in the air. By the time Operation Roundup was completed in early February, UN forces had reached the Han River and re-captured Wonju.

The Chinese struck back in mid-February with their Fourth Phase Offensive from Hoengsong in the center. A short but desperate siege broke up the offensive in this action, thanks to units of the US 2nd Infantry Division and the French Battalion,

Operation Roundup was followed with Operation Killer in the last two weeks of February, 1951. A revitalized Eighth Army, restored by Ridgway to fighting trim, undertook a full-scale offensive across the front.

It was again staged to maximize firepower with the aim of destroying as much of the Chinese and North Korean Armies as possible. By the end of Operation Killer, the UN forces had captured Hoengsong and occupied all territory south of the Han.

On March 7, 1951, the Eighth Army pushed forward again in Operation Ripper. On March 14 they expelled the North Korean and

Chinese troops from Seoul. This was the fourth time in one year that the city had changed hands. Seoul was in utter ruins; its prewar population of one and a half million had dropped to 200,000, and there were severe food shortages.

General MacArthur was removed from UN Command by President Truman on April 11, 1951 for insubordination. The new UN Supreme Commander, General Ridgway, managed to regroup UN forces for an effective series of counter-offensives. Command of the Eighth Army passed to General James Van Fleet.

Further attacks slowly drove back the communist forces. Operations Courageous and Tomahawk were a combined ground-and-air assault to trap communist forces between Kaesong and Seoul. UN forces continued to advance until they reached an objective called Line Kansas some ten miles above the 38th parallel.

However, the Chinese were far from beaten. In April, 1951 they launched their Fifth Phase Offensive called the Chinese Spring Offensive. This was a major effort, involving three field armies of up to 700,000 men. However, fierce resistance in battles at the Imjin River and Kapyong blunted its impetus. The Chinese were halted at a defensive line North of Seoul referred to as the No-Name Line.

A further Communist offensive in the East on May 15th also made initial gains, but by May 20th the attack had ground to a halt. The Eighth Army counterattacked and by the end of May had regained Line Kansas.

The decision by UN forces to halt at Line Kansas, just north of the 38th parallel, and not persist in offensive action into North Korea, ushered in the period of stalemate which typified the remainder of the conflict lasting from July, 1951 to July, 1953.

Peace negotiations began at Kaesong on July 10, 1951. The rest of the war involved little territory change but during the lengthy two year long peace negotiations fierce combat continued, with large-scale bombing of North Korea.

American losses (killed, wounded, and missing) stood at 75,000 in July, 1951 when the truce talks began. While the negotiations dragged on they would eventually rise to 150,000, including an additional 12,000 dead.

For the South Korean and allied forces, the goal was to avoid loss of any territory and to recapture all of South Korea's land before an agreement was finalized. The Chinese and North Koreans attempted similar operations. Later in the war they undertook operations designed to test the resolve of the UN to continue the conflict.

Principal military engagements during this period were battles such as:

- Bloody Ridge and Heartbreak Ridge in 1951 around the Punchbowl in the east.
- The battle for Old Baldy in the center and the Hook in the west during 1952–53.
- The battle of Hill Eerie in 1952 and the battle for Pork Chop Hill in 1953.

The Peace negotiations went on for two years, first at Kaesong and later at Panmunjom. A major issue of the negotiations was repatriation of POWs. The Communists agreed to voluntary repatriation but only if the majority would return to China or North Korea. Since many refused to be repatriated to Communist North Korea and

China, the war continued until the Communists eventually dropped this issue.

On November 29, 1952, US President-elect Dwight D. Eisenhower fulfilled a campaign promise by going to Korea to find out what could be done to end the stalemate. Eisenhower played his cards close to his chest.

He initiated a build-up of American forces in the region, ordered minor offensive actions, and instructed General Clark to step up the exchange of prisoners with the North. In early April of 1953 the Communist powers signaled they were ready to negotiate in earnest.

With acceptance of India's proposal for a Korean armistice, a cease-fire was established on July 27, 1953, by which time the front line was back to the proximity of the 38th parallel.

Thus, a demilitarized zone known as the DMZ was established around it. To this day, the DMZ runs north of the 38th parallel towards the east, and to the south of it as it travels west. It is presently defended by North Korean troops on one side and by South Korean, American, and UN troops on the other.

North Korea and the United States signed the Armistice Agreement, however, President Syngman Rhee refused to sign for South Korea. The site of the peace talks, Kaesong, the old capital of Korea, was part of South Korea before hostilities broke out. It is currently a special city of the North.

The total number of casualties suffered by all parties involved may never be known. Each country's self-reported casualties were largely based upon troop movements, unit rosters, battle casualty reports, and medical records.

The United States Armed Forces suffered the following:*

- 33,665 Americans killed in action in Korea
- 3,275 Died there from non-hostile causes and actions
- 36,940 Total Americans who gave their lives in Korea
- 8,176 MIA's still reported
- 92,134 Wounded in action
- 1,789,000 Total of Americans who served in the Korean War Theater from June 25, 1950 to July 27, 1953.

South Korea sustained:*

- 1,312,836 Military casualties, including
- 415,004 Dead

Casualties among other United Nations allies:*

- 16,532 Military casualties, including
- 3,094 Dead

Although the estimated Communist casualties were two million, the economic and social damage done to the Korean nation was incalculable.

---

* Source: Office of Secretary of Defense, Washington Headquarters Services, Directorate for Information Operations and Reports (WHS/DIOR); Defense Prisoners of War/Missing in Action Office (DPMO). Data released 10 January 2000.

Chapter Four

OUR HEROES' STORIES IN THEIR OWN WORDS

THE WAR IN KOREA WASN'T A NOISY AFFAIR. IN FACT IT WAS labeled a "police action" and regarded as necessary but not glamorous. No battle hymns stirred America's soul. But most veterans returned home with wounds in their souls that would never heal.

Silently our veterans returned home from the war and melded into the fabric of our society. They turned their lives from the horrors of battle to jobs and family. Years quickly passed. Like most veterans before them, they simply put it out of their minds by not talking about it, even to each other, until they reached their senior years.

The veteran's stories that follow are the thoughts and memories of their military service they wish to share with their children and grandchildren. The stories selected for this volume represent a wide variety of their untold experiences.

Stories continue to arrive from around the country and I am thankful for each one of them. Your story is welcome too. See the invitation included in the Appendix. The goal is to compile additional volumes of THE FORGOTTEN series until all the stories received are published and shared.

> "A veteran is someone who, at one point in his life wrote a blank check Made Payable to 'The United States of America' for an amount of 'up to and including my life.' That is Honor, and there are way too many people in this country who no longer understand it."
>
> —Author Unknown

MSGT DR. CARL T. ROGERS, USAF
Lawrenceville, GA

"The Forgotten Mission of the Forgotten War"

WHEN ALL THE INTERNATIONAL INTRIGUE WAS HAPPENING, I was a 20 year old Cpl. assigned as a Tail Gunner on Captain (later Major) Max G. Thaete's B-29 crew in the 19th Bomb Squadron of the 22nd Bomb Group stationed at March Field in Riverside, California. We were still in the "Brown-Shoe" Air Force, so the Army designations are appropriate.

The 22nd Bomb Group was placed on Alert Status on 1 July, 1950, and we left March Field on our way to the Korean War on the 4th of July. Due to the need to prevent crew fatigue, we landed in Hawaii and spent the night, then we Island-hopped our way across the Pacific Ocean.

We terminated our trip by landing at Kadena Air Base on Okinawa on 9 July, 1950. We were quartered in 8-man tents, sleeping under mosquito nets because of all the terrible diseases mosquitoes carried in that part of the world. The enlisted members of the crew, along with our Crew Chief, slept in one tent, while the Officers of the crew slept on the other side of "Tent City."

We also discovered that we ate in the 'Mess Tent' and had outdoor outhouses completely screened in. Our morning ablutions were done in a trough that ran the length of "Tent City." The lucky few who got there first in the morning had hot water, heated by solar power. The rest of us shaved and washed in cold water.

The same situation applied to the showers located at the terminus of "Tent City." They were out-door showers heated by the sun which afforded all the Mama-sans hired to keep the grass cut short to deter the Habu snakes an unobstructed view of the GI anatomy. The primitive conditions didn't deter us. After all, we had been stationed at Smokey Hill Field in Salina, Kansas living in tarpaper huts with coal-fired pot-belly stoves in each barracks for heat during the winter before transferring to March Field.

We had also experienced the "steam-heated Quonset huts" in England that had a single, solitary steam pipe running around the perimeter of the barracks for heat during the long cold winters of our three-month TDY's.

We became accustomed to the philosophy that, "whatever doesn't kill you just makes you stronger." There wasn't much else for us to do about it. The United States had demobilized after WWII and we had to make do with what was left after that near total demobilization.

After loading a full load of caliber ammunition for each gun turret and manually loading a full load of 40, five hundred pound bombs, we were ready to go. We flew our first combat mission and hit the oil refinery at Wonsan on 13 July, 1950. My crew, under Captain Thaete's leadership, led the 11 plane formation into the target and every other Bombardier dropped their bomb load when our Bombardier, 1st Lieutenant Simon, dropped ours.

As the Tail Gunner, I had a perfect view of every plane in our formation as well as a perfect view of the bomb-strike on our target area. Smoke and flames billowed up covering the target area, so I knew we had made a perfect strike and I was justifiably proud of my crew.

My combat crew. First Korea Tour: 1950. Left to right: Tail Gunner: Carl Rogers. Left Gunner: Pete Madden, CFC Gunner: Willie Wright, Radar: Major Bosolav, Bombardier: 1st Lt Simon, Radio: Bob Melindy, Flt Engineer: Sgt Bivins, Co-Pilot: unknown, AC: Major Max G. Thaete. Missing is Right Gunner Ken Duncan.

Since this was my very first combat mission and since my only previous combat experience had come from WWII movies, I had expected to see dozens of enemy fighters as well as heavy, accurate flak. Fortunately, there were no fighters and the only flak they threw up was light and inaccurate, which made me quite happy.

I had managed to put on my flak suit in the confined space of the Tail Gunner's compartment — not an easy task — but I did it just this once. For the rest of the missions we flew, I joined the more experienced members of my crew and sat on the flak suit protecting the most essential portions of my anatomy.

After we returned from our TDY to England, in early 1950, my crew and I were sent to McDill Field in Tampa, Florida to participate in a Bombing and Gunnery competition and a Lead Crew evaluation process. After successfully completing the process, we were officially designated a "Lead Crew." There were 27 missions flown on this tour. My crew led every multiple plane formation including the "Forgotten Mission" of this war.

In a Letter of Commendation dated 24 October, 1950, General Douglas MacArthur congratulated the 22nd Bomb Group on our versatility. "When strategic targets were no longer available, I was impressed with the demonstration of the versatility of your crews and their response as you employed them on missions of close support to the Ground Forces, missions which are not normally assigned to medium bombers."

However, despite what Gen. MacArthur had said, we demonstrated our versatility early in the war. On our third and fourth missions, after striking Wonsan and Seoul, we were assigned the task of knocking out railroad and highway bridges to slow the advance of the North Korean troops and to cut their lines of supply.

The advance of the North Korean forces was rapid against the ill-prepared occupation forces and the unsuspecting South Korean forces. Therefore, it was deemed essential to destroy as much of the nation's infrastructure as soon as possible to slow their advance and to prevent a re-supply of the North Korean troops. In so doing, we provided our ground troops the opportunity to reconsolidate their lines and, hopefully, stop the Communist advance into South Korea.

Neither we nor our allied military forces could stop the Communist advance despite our best efforts of bombing the bridges and railroad marshalling yards. Even bombing their capital city of Pyongyang twice had no effect on them. Finally, our ground forces were reduced to holding on to the Pusan Perimeter, just east of the Naktong River in southeastern South Korea along the South China Sea. The situation was desperate.

Back on Okinawa we speculated whether our ground forces could hold on, or whether an evacuation of troops such as occurred at Dunkirk in WWII could be attempted. The defeat of our ground forces and the loss of the Korean War seemed to be inevitable.

Rogers repairing 50 caliber machine gun for B-29 in background.

Then we were briefed that we would mount a large, multiple-unit attack on the Communist forces massed just across the Naktong River near the Taegu-Taejob Corridor. The Communist (North Korean) military forces were poised in mass preparing to drive our infantry forces into the South China Sea and celebrate their victory over the United States and the United Nations forces.

We flew that mission on 16 August, 1950. We were one of the ninety-nine B-29's from the 19th and 22nd Bomb Wings stationed on Okinawa plus one B-29 Bomb Group stationed in Japan. Ninety nine B-29's attacking the enemy's ground forces in what the Stars and Stripes (Pacific Division) newspaper called, "the most concentrated air-strike in direct support of ground forces since the Normandy invasion."

Some of us came into the target area at a mere 5,000 feet and the mission was a total, unqualified success. Quoting the Stars and Stripes again, the reporter accurately stated that our mission was to, "neutralize or eliminate the strength of four North Korean Divisions and several armored regiments concentrated for a full-scale assault against the thin Allied Infantry lines defending the Taegu area," which we succeeded in doing.

The 1,089 B-29 crew members who participated in this mission dropped about 1,000 tons of bombs and, according to the Stars and Stripes, "killed 40,000 of the enemy." This single mission successfully changed the entire course of the Korean War — for a time.

Shortly after this "Forgotten Mission" Gen. MacArthur made his successful invasion at Inchon. The reporters and photographers who normally accompanied Gen. MacArthur rhapsodized over his military genius because, "he marched ashore meeting no resistance." If anyone had asked, I could have told them why.

The enemy who could have and would have resisted the Inchon Invasion were lying dead on the western shore of the Naktong River, their weaponry crushed. The will and the ability of those who survived the torrent of bombs to resist the Allied ground forces were dealt a mortal blow. Their immediate command structure was also destroyed in this "Forgotten Mission."

In 1992, my wife and I took advantage of a fabulous offer from the South Korean government. They said that if a Korean War veteran flew to Korea via Korean Airlines, the government would pay for a six day tour of South Korea, including our hotel, transportation and food. It was a deal too good to pass up, so we signed up for the tour, even extending it to include a three day stay in Hong Kong and a ten day tour of China.

While visiting South Korea, we discovered how grateful the South Korean people were for our efforts to keep them free of Communism's grasp. As we rode from one historical site to another in a bus sporting a sign written in Korean as well as in English, "Korean War Veterans", men would drive their cars close to the bus, honk their horns to get our attention and give us a great big smile along with a thumb's up signal.

Teenagers seeing our bus would blow kisses to us and men and women would wave at us as we rode by with a smile on their faces. It was hard not to get a lump in one's throat knowing how much the common, ordinary people appreciated what we had done for them.

One day we all visited a War Museum where a group of Korean Kindergartners and First Graders were also visiting. I didn't know 'till then that Korean school children learn to speak English in Kindergarten and all through elementary school. They saw us, smiled, and said, "Hello." I bowed and said "Anna-ah-sai-yo" (a salutation). They were thrilled to see us and insisted that we take a class picture with them. Naturally, we all agreed.

TSGT WILLIAM A. "BILL" COWAN, USAF
Saginaw, TX

"Korea, A Real Bummer"

WHO, WHAT, WHERE, WHEN, WHY AND HOW ABOUT LIFE IN the 13th Bomb Squadron, The Devil's Own Grim Reapers, during the Korean Fracas of 1950-1951. OK, you've got your intrepid Gunner (that's me) down at Marine Corps Air Station Iwakuni, Royal Australian Air Force, Honshu, Japan. We're just sort of renting the flight line from the Aussies to park our Invaders.

Along with our sister squadron, "The Friendly Eighth," we will fly our missions to and from Korea from this former Japanese Naval Air Force Base. Fortunately, little damage had been done to this rather large land and seaplane base prior to the surrender of the Japanese military.

Therefore, most of the original Japanese facility was usable including barracks and administrative buildings. Only a few minor areas of the base had been strafed by visiting Allied fighters prior to the surrender.

As previously noted, this area on Southern Honshu Island was turned over to the various British Forces under the Occupation Agreement. All was calm and serene prior to June, 1950 when the Bad Guys from the North crossed over into the Good Guys territory south of the 38th parallel.

While stationed with the 6132d TacAir Group at Pusan, we had an unpleasant neighbor located just to our south. This was the fledgling United Nations Cemetery that was unfortunately increasingly busy. Seems one of the Russian pilots had been shot down over an area controlled by us, and his body was recovered.

The UN Command attempted to return his body to the Russians but they steadfastly denied his identity. To preclude interring an enemy combatant in our sacred ground, a plot was opened several hundred yards from the UN grounds, and he was buried with full military honors. Whether or not the Ruskies recovered his remains after the armistice is unknown to me.

As previously noted, the early days of the Korean fracas were hectic to say the least and the men of the 3d Group were no exception. Pilots and planes were sorely pressed to defend against the hordes from the north. Every available asset was pressed into service and fatigue was a common factor for planes and personnel.

Munitions were in extremely short supply and bomb loads were eclectic for lack of a better word! The "kitchen sink" was eyed by the Armament Troops on more than one occasion. Bombs were being flown in daily from stockpiles on the island of Okinawa and immediately loaded into the waiting aircraft. These things and history were gleaned from the early operational reports, since I was still playing around with my jeep in the 6132d TacAir Group on the other end of the train.

Cowan — "Cool Hand Luke"

During the winter of 1950, the situation settled down to some degree. Additional aircraft and crews were arriving to bolster the two squadrons of the 3d Group who had been flying constant sorties daily.

Now I was qualified to sit back in the gunner's compartment of the lethal Invader. As long as I draw a sober breath I'll not forget my first combat mission. It was cheerfully called The Dollar Ride. Lieutenant Prettyman was the pilot, and my friend. I'll have to admit that sucker scared the britches off me.

Our assigned area was in far northwest Korea in an area adjacent to the Chinese/Russian border. I do believe the good Lieutenant had a death wish! Later, while training new crews at Langley AFB, VA, he pulled the wings off an Invader over Chesapeake Bay showing the proper strafing procedures.

He flew directly toward a border outpost with his landing lights on hoping they'd fire at us so he could return the compliment. Failing at that ruse, he turned our attention to legitimate targets in the area.

Now remember, this was my first combat mission sitting in the gunner's compartment located aft of the bomb bay and separated from the cockpit area by that bay. The intercom was my only link to the folks up front driving this beast. We had completed several firing passes on ground targets when "old hotshot" decided to knock down a smoke stack on the local power plant.

The only major problem was, he didn't let me in on the deal. So sitting back in my cubby hole, I was attuned to possible ground fire I might be called upon to suppress. As we started our firing pass, and without so much as a howdy-do to me, he unleashed a salvo of ten five-inch HVAR rockets at the stack.

Now those rocket motors out on the wing were right next to my ears, and the Gates of Hades opened when he fired. I immediately grabbed my parachute and prepared to leave that mortally wounded aircraft. I just knew we had taken a direct hit from ground fire of some sort. To put it mildly, I was scared out of my gourd!

Other times this sight came into its own was when we were outbound from late sorties with early dawn light. One mission will always rank high in my memories. I'll quote from the citation awarding me The Distinguished Flying Cross for combat action which is rarely awarded to enlisted gunners. It is more often awarded to the rated commissioned pilots.

> "Staff Sergeant William A. Cowan, Jr. distinguished himself by extraordinary achievement while in aerial flight as a gunner of a B-26 bomber over North Korea. On the night of 18 April, 1951, while patrolling road and rail supply routes south of Sinanju, Sergeant Cowan discovered three trains in a marshaling yard.
>
> "Upon initiating an attack, his aircraft encountered intense anti-aircraft fire from six twenty millimeter batteries located on surrounding hills. Only through Sergeant Cowan's withering counter fire, which held down the enemy gun crews, was an attack possible. So accurate was his fire that all batteries were silenced and he was able to turn his guns upon the primary target.
>
> "He ably assisted the pilot in destroying one train and damaging two others before all of the pilot's ammunition was expended. Departing the scene of the attack, Sergeant Cowan's aircraft headed toward base and soon sighted another enemy train.
>
> "With only the guns of his upper turret firing, Sergeant Cowan poured fifty caliber bullets into the train's entire length destroying the locomotive and ten box cars while damaging every section. By his accurate marksmanship, Sergeant Cowan dealt a material blow to the enemy transportation system."

When you fly the way we did, down low and dirty, the mistakes can really ruin your whole day. It usually resulted in an empty spot on the aircraft ramp, and empty cots in the barracks. First a toast

Gunner Bill — Scourge of Anti-Aircraft Positions. Upper Gun Turret of B-26 Bomber. Bill was Awarded The Distinguished Flying Cross

and prayer to the memory of our fallen comrades; then on to briefing with two more missions under our belts. Night falls over Korea!

Korea was a strange war experience for those of us who had never experienced fighting against those of oriental extraction. Having served three years in Japan, I thought I had some understanding of the oriental mind.

However, while fighting the Japanese in the Far East during WWII our forces that found out they had a different mind set and all the "newcomers" had a serious learning lesson. Some examples of this mind set were; their mass frontal assault in broad daylight against a well entrenched enemy; the utilization of second rate warriors, armed only with grenades, to breach the defenses where they would retrieve discarded weapons with which to fight; and night attacks in force with bugle blowing and electronic sound effects.

Couple that set of tactics with the early wartime experience of our front line troops. They were issued only a set amount of ammunition per weapon/ man, according to outdated Army regulations called a Table Of Allowances, that were still in effect and being adhered to by our zealous leadership. It was almost criminal to a suicidal degree against our forces. Check it out, it's true. I was there when it happened in 1950.

SGT DONALD BAUMANN, USMC
Cocoa Beach, FL

"Profile of a Marine"

Submitted by Rick Kennedy, Secretary
1st Marine Division Association, Central Florida Chapter

DON BAUMANN JOINED THE UNITED STATES MARINE CORPS from his home in Miami, Florida at age 17, but didn't tell anybody that he served in the United Sates Navy Reserves in 1947 as a 15 year old lad who fibbed about his age. He served in the Navy Reserves training as a flight engineer aboard a PBY sea plane at the Naval Air Station in Opa Locka, Florida.

This experienced veteran had some concerns about the transit barracks at Yamassee, South Carolina, but it was smooth sailing for Don at Parris Island. Afterwards, he was sent to Camp Lejeune, North Carolina for advanced combat training, but this was called off at the advent of the Korean War.

Then it was off to Camp Pendleton, and soon aboard the General Meigs from San Diego to be assigned to the 7th Motor Transport Battalion of the 1st Marine Division. On September 15th, Don saw his first combat action on the landing at Inchon, and he marveled at the sight of 16" shells lighting up the sky overhead as they were fired from the great battleship, The Missouri.

Later, during the assault on Seoul, Don helped direct his convoy of trucks during a night operation. They were loaded with pontoon bridge equipment that was needed to cross the Hun River, since the bridge had been completely destroyed by the enemy. They passed by the road to Kimpo, where the bridge was to be off loaded, and all the trucks ended up on the blown-up bridge over the Hun River.

Don helped direct a turn around. After the other trucks departed, he began turning his truck around and got it stuck in a ditch. He was concerned about leaving the truck, and thought about destroying it with a phosphorus grenade to keep it from being used by the enemy. In a last minute attempt, he started the motor and miraculously succeeded in moving the truck out of the ditch. He later joined his unit, which had come looking for him, and led him back to Kimpo without a scratch.

After the victory at Inchon, Don joined parts of the 7th Marine Regiment and boarded an APA to skirt the southern perimeter of Korea and made a landing at Wonsan. From there the trucks of the 7th Motor Transport battalion carried the supplies of the 5th, 7th, and 11th Marines to Hungnam. Then they moved up through Koto-ri, to Hagaru-ri, and to far away Yudam-ni above the Chosin Reservoir where he could see all the way to Manchuria.

Soon after Thanksgiving, the temperature got down to 40 degrees below zero, and the Chinese Army came at the Marines in great numbers. Don's truck drivers became Marine riflemen and they joined the 7th Marines to fight off the on-coming enemy that outnumbered the Marines by twelve to one.

Soon the bridge at Yudam-ni was also destroyed, and the US Air Force dropped bridge spans to allow the 1st Marine Division to make a crossing and then fight their way to Hungnam.

It was indeed a tough road advancing to the rear under extremely adverse weather conditions. They carried the dead, wounded, and walking Marines back down through Hagaru-ri, then to Koto-ri to the awaiting ships of the US Navy at Hungnam where the Great Battleship Missouri was shelling the advancing Chinese Army. The 1st Marine Division and all attached units were awarded three PUC decorations for their actions in these campaigns.

Don joined the 1st Marine Division at Masan in early December, and with replacements coming to South Korea, they began preparing for Operation Killer during the early part of 1951.

Soon after, Don was relieved from Korea on the point system, and he was sent to Cherry Point, North Carolina where he was positioning himself for flight training. He became Sergeant of the Guard, but passed out one day on duty. He was sent to the hospital and they diagnosed his illness as Malaria. While recovering he was assigned light duty as a drill instructor for the Women Marines.

One day while drilling the women Marines in close order formation, he was stopped by General Cooley, Commander of the 3rd Marine Air Wing. The General got out of his car and told Don that

Left: Marine Baumann; and (right) with prisoners at Chosin Reservoir.

those women Marines were one of the finest marching platoons he had ever seen.

General Cooley then made arrangements for Don to keep his dream and attend flight school at Pensacola, Florida, but Don's family was in distress and they needed Don to be at home, so he ended his Marine Corps career.

Don met his lovely wife Margaret in 1952, while serving with General Cooley in Miami, FL and attached to Headquarters Squadron 3rd Marine Air Wing. They have been married more than 52 years; have five children, 11 grandchildren, and two great grandchildren,

Don spent his career after the Marine Corps with Pan American Airline and retired at the space center for NASA where he was a method analyst. The Astronauts gave Don a "Silver Snoopy Award" and pin that had been flown in space. The award was given for *"Exceeding professional dedication and outstanding support that greatly enhances flight safety and mission success."*

It is indeed a great honor to participate in the meetings and functions of the Central Florida Chapter of the 1st Marine Division because of the high caliber of members like Don Baumann. He participated and survived in one of the truly great battles in the history of the United States Marine Corps.

LT STANLEY GROGAN, JR., USAF
Pinole, CA

"First Airborne Toast of the Day"

I AM AN AIR FORCE RESERVE LIEUTENANT WHO SERVED WITH the 68th "Lightning Lancers," users of the Lockheed F-94 Starfire all-weather fighters. I flew combat with them on 20 night missions.

Left: Pilot Aubill and Grogan on F-94B; Right: Grogan, 2nd tour

This outfit was the first to down an enemy aircraft in combat, first to be night fighters in Korea, first to begin interdiction, first to escort evacuees from the Korean Peninsula, and first to use the all-weather jet fighter in combat.

The business of flying Lockheed F-94 Starfires around-the-clock in defense of Southern Korea was grim work demanding utmost concentration while boring through the murk after a "bogie." Sometimes an incident would occur, which made the life of all-weather flying seem very normal.

At dawn one chilly morning, Major Rogers D. Littlejohn, USAF, formerly operations officer of the famed 68th "Lightning Lancers," appeared at the Itazuke alert shack attired in flying gear. He picked me as radar observer and sat in the shack awaiting the "scramble" call. Suddenly, the warbler whipped us to attention.

"Scramble Red, vector X-ray," took us into our jet, "Leaping Lena," on the double. Rogers wound up the engine and snapped the canopy shut. As we pulled out on the active runway, the sleep of all was cracked by the roar of afterburners sending us airborne.

We settled out on course, noting an ominous tone in the commands of the ground controller: "Bogie 11 o'clock, 16 miles, angels unknown." I thought it was in the vicinity of the Straits of Tsushima, and judging from the direction of approach, it could have been anything.

Several minutes later we closed to zero distance between us and the bogie, with control noting that "the blips have merged." This was great, I thought. We had missed a tactical intercept. As we began a 360-degree turn for a visual check, Rogers spotted it on the deck. "We've got it, Frogman (control's call sign). We're going down to investigate."

As we crept up on the tail of the flying boat, I could hear Rogers charge the guns as I prepared to read close range distances for accurate firing if necessary. Then we pulled slightly to the starboard of the tail and noted the British insignia. Whew! No D-day this!

Flying formation with it for a few seconds, Rogers and I heard this interesting VHF message: "Why don't ye join us for tea, Yanks"? At this point, the entire crew of the flying boat showed their faces and full tea cups at the ports of the plane in the first airborne toast of the day.

We waggled our wings in return and did a peel-up to the right on course to Itazuke with afterburner on just to show 'em we were in there pitching, even if we couldn't drink tea while intercepting.

My current hometown, for the past 46 years, is Pinole, CA, the oldest incorporated city in the state of California. Upon my honorable discharge from the USAF in 1953, I was employed by the Central Intelligence Agency.

Subsequently, I was recalled into the USAF in 1957 and served as a radar intercept officer in the F-89J. In 1993 my family and I returned to Korea where I received the Medal of Peace. I recalled writing to the President of Korea that I had never seen Korea in daylight since my two combat tours in the F-94B and RB-29A were all at night.

CPL BRYAN 'Jerry' WALKER, US ARMY
Kansas City, MO

I SPENT ABOUT NINE MONTHS IN KOREA WITH A LINE CO. BEGINning September 15, 1951. I only had a couple of close calls, but how I got there was [almost] a laughable misuse of Military Occupation Speciality (MOS).

I had about eight weeks of basic training with a carbine at Camp Chaffee, AK. Then to Fort Sill, OK, for eight to ten weeks of 105 Artillery Fire-Direction, before going to Japan. They gave me four rounds of ammunition to sight in an M-1 before shipping me out to Korea. I had only fired ten rounds with a Garand at Camp Chaffee.

They changed my orders to go to Eta-Jima for four weeks of climbing telephone poles, and finally shipped me out to Pusan, Korea. Then the telephone poles got smaller and smaller. The replacement depot changed my pole lineman MOS to field wireman, and I ended up in Co. C. 7th Regiment 3rd Division as a Corporal.

"My Closest Call — or Surviving Stupidity "

Shortly after regaining Hill 355, known as Little Gibraltar, while repairing a sound power line, I got a flat trajectory incoming. I was below the front ridgeline but unaware that I was exposed on the flank. Near miss but a miss! Looking to reclaim my carbine, I discovered both it and the three or four inch stump I had leaned the weapon against, were gone.

There are not too many extra firearms floating around a line company, but a guy in a rifle platoon loaned me a, "How does this thing work" off-breed Czechoslovakian pistol to run a patrol.

Too dumb to celebrate my survival, I was instead starting to worry about a statement of charges for a lost weapon. Then a guy coming up the hill a day or so later says, "Does this belong to anybody up here?" The taped clips were still in place but the stock was long gone.

The Army tried to bill me $3.45 for a lost first-aid pack four or five months after my discharge and I still expect a statement of charges for that damn carbine to arrive any day now. Little Gibraltar, with a couple of prominent peaks soon became "Dagmar" to honor a well-endowed lady celebrity of the 50's!

Aside from a patrol skirmish and a guy in Baker Co. trying to kill me with a grenade after I had installed a listening post line, there wasn't much to do but keep my head down, and count points. Eventually, there came the magic number — 39 points!

I returned to the states and found a welcome surprise. For about a month the powers that be granted me a dream assignment interviewing inductees at Camp Crowder, MO. There were gourmet cooks, a Class A pass, the big-little city of Joplin, MO, just 20 minutes from us, and a chance to go home to Kansas City, MO, on the weekends.

Whoa, wait a second. BIG Mistake! Returnees with a combat MOS must serve in a combat capacity. New orders arrived, and with about 60 days left, I'm living in a Quonset hut on a pier in the middle of Soo Locks, MI, shaving in cold water, and designated as part of a 40mm triple A gun crew where the only man possessing the proper MOS is the crew chief. (Some things never changed)

The communications guys had tapped into a line. They could report the gun ready for action, and monitor the long pier from a strategically placed bunk, without getting out of bed. By the time I left in November of 1952, the Army had forced us to remove the gun barrel at night, and replace it at sunrise to ensure the gun was manned and operable.

Army service is great education and experience, if you escape unscathed. The pay isn't too good! Before being drafted I had a big interest in bikes. I didn't see a lot of action while in Korea. Just enough to think, "I ought to forget those infernal machines and settle down when I got back home."

However, after returning home, I bought another bike during my 30 day leave, before reassignment. Say goodbye, to the best laid plans

Walker — Our typical high-rise while living on the line.

of mice and men. I eventually ended up in business with the guy that sold me the bike while I was on leave, and I became another motorcycle malcontent. By the time I finally sold my shares, old age had finally settled me down. I had an old English machine that ran about 130 mph in the 1950's but that and my draft status have disappeared. Time flies!

SGT CHARLES APPENZELLER, US ARMY
Haines City, FL

I ARRIVED IN SOUTH KOREA IN NOVEMBER 1952 BY TROOP SHIP at Inchon at high tide. They have a real low tide at Inchon. We landed by landing boat and climbed up the seawall with our duffle bags, not bad. We then boarded a train with no windows and left. I got some cold C-rations to eat. We went to the 25th Division and then to the 21AAA and ended up in Baker Battery.

I was sent to track 132 and learned how to work on the Quad 50 machine guns. I replaced a man who was hit at Heartbreak Ridge. We were giving air support to the 69th field with a 155 Howitzer. From there we went to give air support for the 8th field with a 105 Howitzer firing on Old Baldy, Arrowhead, and White Horse. We were fighting for Old Baldy.

The 8th field fired all day and all night and two 155 mm Self Propelled Howitzers came up and started firing. From there we went on the multiple rocket launcher (MLR) giving support to the 1st ROK, South Korean Infantry Div. We fired on Arrowhead Ridge which was being used by the Chinese for patrols.

We had a lot of rain that month which really messed up the dirt roads. When we brought our track down from our position, it was so slick we slid half way down. From there we went to the Panmunjom area and watched the trucks carrying the sick and wounded POW's (Prisoners Of War) back to the rear.

We went to our new track position, which I did not like. We were to work with one of those old Sherman tanks. The tank driver was to go up to his place first and then I was to put my track next to him. The tank driver started up with one man directing him, but it was too steep and gravelly. The tank started to slide side-ways and rolled down. The man was thrown clear but the tank went all the way and landed upside down. The driver was not killed but he was badly hurt.

Appenzeller on Track 132

There was no way I could have made it. The Colonel came and told us that no one goes up there until it is fixed. A bulldozer came the next day and made a new position for our track. I did not want to go back up there in that place.

My track later spent a few days at Pork Chop Hill. We were there in relief of those guys. We had lost that outpost, and Division and 8th Army did not think it was worth more lives at this time. I was there about seven days before being sent to a new location.

We moved to the new concrete bridge and were with the 8th field again. We took a lot of rounds while we were there. The Chinese thought they had us zeroed in but they did not.

The last day of the war we had three rounds come over and hit at the bottom of our position but never exploded. We never went down to see why. I believe this happened on July 27, 1953, and at 10 p.m. that night, the war was over.

CPL WILLIAM O. BRENNEN, USMC
Larkspur, CA

"Angels or Gooney Birds"

ON 25 JUNE 1950, THE NORTH KOREAN MILITARY FORCES launched a massive surprise assault on South Korea. US Army units in Japan were quickly committed as part of the United Nations effort to block the North Korean invasion. On 14 July 1950, the hastily assembled 1st Provisional Marine Brigade left the US for Korea to reinforce embattled US Army and South Korean troops defending the Pusan Perimeter at the southern tip of Korea. The North Korean Army had nearly conquered all of Korea.

I was a Corporal assigned to Headquarters Squadron, Hq. Sq. 33 as a radio/radar technician, and back-up airborne radio operator. Hq. Sq. 33 had a compliment of about eight F4U Corsair fighter/bombers, four F7F photo-recon planes, and two R4D transport aircraft.

Left: Brennen; Right: Brennen with Armed Corsair in 1951

In August and September of 1950, additional Marine air and ground units were sent to Japan to prepare for the Inchon landing set for mid September. The Marine Brigade was absorbed into the 1st Marine Division, and M.A.G. 33 became part of the 1st Marine Air Wing.

The Inchon landing resulted in the defeat of the North Korean Army. In late November and early December 1950, the UN forces were driving toward the Yalu River and the Korean/Chinese boarder. The war was all but over and we expected to be home by Christmas.

At the Chosin Reservoir a force of over 100,000 Chinese Communist soldiers had surprised, encircled, and trapped a force of about 18,000 Americans comprised of the 1st Marine Division, the 31st RCT of the Army 7th Division and 300 British Royal Marine Commandos. The bitterly cold Siberian winter was a disabling enemy to both the UN and Communist forces.

I was assigned as radio operator aboard 50785 on 30 November 1950. This was a twin engine transport and the slang term for the aircraft was "Gooney Bird." Most of this narrative will be devoted to my personal observations while serving in that capacity.

As of 28 November 1950, the crew of 50785 consisted of 1st Lt. Bobby Carter, pilot, 1st Lt. J. Flickinger, co-pilot, M/Sgt. John Hart, crew chief, and S/Sgt. Arthur Allison, radio operator. The crew of 12436 consisted of M/Sgt. Robert Brown, pilot, Capt. W. Van Ness, Co-pilot, T/Sgt. James Morris, crew chief, and Cpl. Ed Farra, radio operator.

Marine ground forces were divided into three basic units. The First Marines were at Koto-ri, 11 miles south of Hagaru-ri, the village at the Southern tip of the Chosin Reservoir. The Fifth and Seventh

Marines were at Yudam-ni, 14 miles North of Hagaru-ri. Division Hq. was located at Hagaru-ri, and it was garrisoned by mostly service troops and various detached units from the three regiments and the 11th Marines.

A narrow, winding, mountain road, called the Main Supply Route (MSR) connected the three forces. The Chinese had cut the MSR between the three units, which isolated each of them.

Toktong Pass was a vital position between Hagaru-ri and Yudam-ni, a key part of the main supply route connecting these two villages. The 240 Marines of F/2/7 were assigned the job of holding the pass open. On 27 November 1950, Chinese forces began an assault on the pass that was to last several days. F/2/7 held, but at great cost.

They were finally relieved on 2 December by Marines withdrawing from Yudam-ni to link up with the Marines at Hagaru-ri in order to break out of the Chinese trap. F2/7's Commanding Officer, Capt. William Barber, was awarded the Medal of Honor for his heroic stand at Toktong Pass. His citation reads in part,

> "Capt. Barber took position with his battle-weary troops and, before nightfall, had dug in and set up a defense along the frozen, snow-covered hillside when a force of estimated regimental strength savagely attacked him during the night, inflicting heavy casualties and finally surrounding his position following a bitterly fought seven hour conflict.
>
> "Capt. Barber, after repulsing the enemy, gave assurance that he could hold if supplied by air drops, and requested permission to stand fast when orders were received by radio to fight his way back to a relieving force after two reinforcing units had been driven back under fierce resistance in their attempts to reach the isolated troops."

On 28 and 29 November 1950, the crew of 50785 made para-supply drops to the surrounded Marines at Toktong Pass, as requested by Capt. Barber. The drops were made at very low altitude to ensure that the supplies went to the Marines, and not the Chinese, as was the case with many of the supplies dropped by larger aircraft that made their drops from a higher, safer altitude.

Ground fire was very heavy on the second day, and 50785 took several hits. S/Sgt. Allison, the radio operator, was seriously wounded when the aircraft made its second run over the drop zone. He had been shot through both legs and lay bleeding on the floor of the aircraft. M/Sgt. Hart had to apply tourniquets to both of Allison's legs to keep him from bleeding to death. After stabilizing him, we made additional low altitude runs over the drop zone to deliver the rest of its cargo. The crew risked sacrificing themselves to ensure the re-supply of the embattled Marines on the ground.

On one of our evacuation flights, we were orbiting above Hagaru-ri, waiting for an Air Force C-47 to take off before we could land. We observed the Air Force plane crash-land just after taking off several hundred yards beyond the end of the airstrip. It did not catch fire and the occupants got out of the plane and staggered back toward the airstrip. They were in Chinese territory, but the Chinese made no effort to intervene, that I could observe.

After we had landed, off loaded, and were taking wounded aboard, a small group of beat up looking Marines hobbled up to MSgt. Hart and the Senior NCO stated, "We've been surrounded up here for days. We're half frozen. We've all been wounded, and we just survived an airplane crash. For the glory of God and the Marine Corps, get us out of here!" We accommodated them.

In early 1998, I saw an article in the Travis Air Museum News by Walter Kane, USAF, retired, entitled "Last Flight to Hagaru-ri." Mr. Kane was the engineer aboard the plane I saw crash, and his article described the incident I had witnessed. In his story, Mr. Kane stated that after making it back to the air strip following the crash, they boarded another C-47 for the flight out, preferring it to another Marine transport that had 20 or 30 bullet holes in its fuselage.

I'm convinced the Marine plane he referred to was 50785. Mr. Kane lives close by so I contacted him and we have had several enjoyable visits, reminiscing about our mutual experience 50 years earlier in Korea.

Hagaru-ri Air Strip, Chosin Reservoir, Dec. 1950

On 9 December, the crew of 12436 manned 50785, and made two evacuation flights from Koto-ri, bringing out an additional 50 to 60 wounded troops. Crew members, TSgt. Jim Morris and Cpl Ed Farra, told me that Chinese troops were entrenched within 200 yards of the air strip at Koto-ri, their positions clearly visible. They did not fire on the transport planes because of strong Marine defensive forces opposing them. The two sides remained at a standoff.

The crew of 12436 brought out Marguerite Higgins, a correspondent for the New York Herald Tribune. She was the only female war correspondent in Korea, and the most famous of the Korean War. In 1951, she was awarded a Pulitzer Prize for her work in Korea.

When the 1st. Marine Division and accompanying Army units left Hagaru-ri, and later Koto-ri, they were adequately re-supplied, and they were not encumbered by wounded. They were able to devote all of their resources to breaking through the Chinese Communist armies that tried to block their withdrawal to the sea.

The ground crews worked long hours under miserable conditions on the wind swept, snow covered airfields at Yonpo and Wonson and the ice encrusted flight decks of the aircraft carriers at sea, in order to keep the fighter/bombers and transports in the air. All of us had but one objective: to do the best job possible at whatever task we were assigned, in order to get our brothers back out of those mountains.

In nine days of hectic flying, over 273 tons of supplies were delivered, and 4,690 wounded troops evacuated from the two primitive air strips at the Hagaru-ri and Koto-ri. Even though two transport aircraft were lost during the operation, none of the crashes resulted in a fatality. All of the wounded were delivered safely to Yonpo Air Field for relay to a nearby hospital ship or airlifted to hospitals in Japan.

I don't know any man who would not have willingly laid down his wrench or screw driver, picked up his M1 rifle, and headed up the road to Koto-ri if he had been asked. In fact, one of our mechanics did just that. He obtained permission for us to fly him into Hagaru-ri, where he spent the night on the perimeter with his brother who was assigned to an infantry unit there. We flew him out the next day.

I know that 50785, her sister aircraft 12436, and the ground crews that kept them in the air, were most certainly a deciding factor in the success of the "attack to the rear." I also suspect that all of the wounded troops evacuated by the C-47s and R4Ds, if given the chance, would vote to change the nickname of those aircraft from "Gooney Bird" to "Angel".

Top: Marine 50767, Chosin Reservoir. Crash-landed 2 Dec. 1950, destroyed 6 Dec. 1950. Pilot: Capt Paul Noel, Crew: TSgt D. Schwitzer, Sgt K. MacLeod; Bottom: USAF C47 being off-loaded at Hagaru-ri Air Strip, Chosin Reservoir, Dec. 1950. The dead, shrouded in sleeping bags (at lower left), await evacuation.

TSGT DALE L. HANEN, USAF
Brunswick, GA

"My life with the 5th Mule Train"
or
"Jackasses of Korea"

O/S project 0211, /S Asgmt on Project "Able Charlie Dash Baker."

I WAS GOING TO EMORY RIDDLE SCHOOL OF AVIATION IN OPA-Locka, FL, but I washed out after three months and was transferred to the motor pool at Kessler AFB in MS.

I came back off leave in November, 1951 and saw a sign-up sheet in the day room of the Motor Vehicle Squadron at Kessler AFB asking for people to go to the 5th Motor Transportation Squadron in Korea. Being young, I signed up! When I got to Korea, I went to flight operations to see about a ride over to Young Dung Po — which I could not pronounce.

There were no rides and no phones. The man on the desk told me he had just arrived on the same ship. The next two guys, an Air Police and a driver, had just arrived there also. When we got to the main gate, the Air Police stopped a truck with three black guys in it. They said, "We know where the Mule Train is!" So after I climbed into the back they let the tarp down, and we were off.

When I got there and went into the orderly room; The CQ, from Minnesota, gave me his tray and stuff to go eat. Well, suddenly there was an air raid, so I ate in a foxhole! I stayed a few days and went down country to Wonju, K46.

In all the time I was in-country, I never met the First Sergeant or Commander at headquarters. The Commanding Officer was a big fat Major who wore his .45 on his middle. That is the only place it would stay. I was only at HQ a few days, and most times just overnight.

At Wonju, we were on the same side of the base as the gas dump, munitions, bombs, 50 calibers, napalm tanks, and powder. The mess hall, showers and everything else on the base was on the other side of the runway. The P51's were parked with their guns facing us.

We had all tents except the maintenance shop, which was made out of anything and everything. The detachment commander had his office in the orderly room, with a wall in between. His tent quarters were about 100 yards away.

One day we had five minutes for a break, and I ate frozen spaghetti right out of the can for supper. We didn't haul rations that I can remember, except for ourselves. We did haul beer one time. They stopped us after a few cases were missing from the loads!

Suicide Mountain! K18. I can remember it just as if I was driving it today, or was looking at it today. We never loaded at night, and very seldom kept loaded trucks in our area. The Army would come in with loaded trucks and pay off the first guard to watch out for the items on the truck. We took only what we needed; nothing to sell or trade! We could easily get the Army or Marines to trade with us, if we had booze.

We hauled gas, bombs, 50 calibers, napalm, tanks and more. We kept the trucks in first gear to go down and 3rd gear to go up if we could. When we got to Kangnung, we would sleep in Quonset Huts. This is the only place we slept in a hut. Everywhere else it was tents.

I was coming back to the base one day and three fighter pilots stopped me and wanted a ride back from town. When I stopped to give them a ride, they took one look at my load and refused to ride with me. I had a full truck load of 500 pound bombs.

We would get to the bomb dump and ask in which area to unload. Then we'd back it up to drop our tailgate. Then we'd go backwards real fast, hit the brakes, and let them roll off! I found out about 15 years later in Thailand, that they do explode. It made a big noise too! We lost one man.

The first six months we could not get anything to drink, but we would get a beer or two in the chow hall once in a while. Then we opened an E-club, so we could get something there.

I left Korea February 3, 1953 to go to Japan for a ship home. I arrived back in the states and had to stay a week at YBI (Yuerba Buena Island/ Treasure Island Naval Station) San Francisco. After leaving Minneapolis, I went to Chanute AFB in Illinois. I met my wife, who was a GI brat, while I was driving the base bus. We were married June 29, 1954.

Hanen with the 5th Mule Train in Korea

GYSGT HAROLD L. GROVES, USMC
Jackson, TN

I ENLISTED AS A PRIVATE AT THE US MARINE CORPS RECRUITING Station, Indianapolis, IN on 10 February 1951 for a three year hitch. They sent me by train to MCRD, Parris Island, SC. I arrived on 12 February 1951 and began recruit training. After graduating in April of 1951, I went home on leave for 10 days.

I was assigned to casual company after returning to MCRD, Parris Island, SC. I was then transferred on 24 May 1951 to Camp Lejeune, NC and assigned to First School Co. Engineering School BNMB.

While waiting for travel orders to Korea, I was promoted to Cpl on 23 November 1951. I attended cold weather training at Pickles Meadow's Training Center, located in the Sierra Mountains north of Bishop, CA. I arrived Thanksgiving Day and remember eating cold turkey and dressing out of my mess kit.

On 14 December 1951 we boarded USNS General William Weigel T-AP-119 in San Diego, CA. We departed the next day and disembarked at Kobe, Japan, on 31 December 1951 in the middle of a terrific storm.

We boarded a Japanese train and went to Sasebo, Japan. When we arrived, we were transferred by truck to an Army Camp named, Camp

Sgt. Groves with camp dog

Mower. After a couple of days of waiting, we were transported by truck to the pier at Sasebo, Japan, and boarded a Japanese ship, Red Ball Liner.

We departed from there on 3 January 1952 and arrived and disembarked at Pusan, Korea, the next day. We were met by the US Marine Corps trucks at this point, and taken to our new home at Masan, Korea.

I was assigned guard duty for 60 days and served as Corporal of the Guard. After this, I was assigned to the Heavy Engineering Equipment Sections and given duties as a tractor equipment operator, a road grader operator, and crane operator unloading incoming supplies from barges onto the dock. I did mostly road work, maintaining roads in that area.

I helped build a prisoner of war camp for North Korean prisoners. I would go out on the roads alone on my road grader and work all day. The only protection I had was my M-1 rifle. It was locked and loaded by my side on the seat of the road grader. I would stop at Army camps along the way and get permission to have chow with them. They were always very nice and appreciated the work I was doing on the roads.

I was promoted to Sergeant on 1 June 1952. It was a long hot summer and I thought it would ever end. Winter finally came and I knew it wouldn't be long before I would be heading home.

When the New Year came, I was still in Korea. On 10 January 1953, I boarded the USS General M. C. Meigs at Inchon, Korea. I departed from there on 10 January 1953 and arrived at San Francisco, CA on 26 January 1953. That concluded my first tour in Korea.

CPL JOHN RICK KENNEDY, USMC
Port Orange, FL

Submitted by Rick Kennedy, Secretary,
Central Florida Chapter, 1st Marine Division Association

IN THE SUMMER OF 1950 I WAS ATTENDING SUMMER SCHOOL at the Indiana University extension in Jeffersonville, IN. After class in July, I joined some friends at the "Brown Derby" bar who, like me, had served in the military at the end of WWII. On the television screen the Marines were making a landing in Korea. I told my buddies that the war would be over in two weeks. They said, "Rick if you think so much about the Marines, you should go over and give them a hand."

The next day I joined a reserve outfit in Louisville, KY, and two weeks later headed by troop train for Camp Pendleton, CA to receive advanced combat training. I was trained at tent camp as a machine

Left: Kennedy; Right: "Saddle-up and move-out"

gunner, and then introduced to the 75 recoilless rifle mounted on a heavy tripod.

In late October 1950, on the Third Replacement Draft, I headed for Korea aboard the APA General Collins. We landed in Kobe, Japan and traveled to Otsu to pick up our winter clothing. I was very happy to be assigned to the Fifth Regiment. I had written a history on the Marine Corps for college, and was well aware that this was the premier Marine Regiment.

We landed in Pusan, South Korea, in early December 1950. Then we traveled by train to the bean patch at Masan. We waited there for the 1st Marine Division to return from the Chosin Reservoir. I had been assigned to Charlie Company of the 1st Battalion, 5th Marine Regiment. The men returning were haggard and dirty and looked like the homeless on Clark Street in Chicago, IL.

We enjoyed turkey for Christmas, and in no time we were looking like a first class regiment and ready to roll. Soon we left Masan with full packs aboard LST 104 that was built in Jeffersonville, IN. We spent two days at sea, and landed in Pohang, on the eastern shore of South Korea.

Our company objective was to locate a regiment of North Korean guerrillas that had broken through our lines and were causing family destruction while ravaging small towns for food. We left Topyong Dong after dark, and headed for the village of Chachon-Dong in a forced march that made boot camp seem like child's play.

We arrived before sun up and moved into the homes of the Koreans. This was a planned ambush so we only went outside after dark. After several peaceful days, I was awakened by machine gun fire and the party started. It did not last long and the enemy disbursed with dispatch.

Early in February 1951, we moved north by truck to Chungju to be the lead company in Operation Killer. This paved the way for the 1st Marine Division and all elements on the front line to eventually advance to north of the 38th Parallel.

On February 21st 1951, we assembled near General Douglas MacArthur who was seated in his jeep. We attended Mass in the early morning and then jumped off on the attack by sliding 100 feet down the side of a snow covered mountain.

The first night in my fox hole, my sleep was disturbed by the sound of dueling artillery barrages. For the next three months we advanced in the mountainous central section of South Korea. We lived in the dirt in freezing temperatures after digging new fox holes most every night. We climbed mountains and confronted the enemy most every day. My jobs were rifleman, BAR man, runner, and later as a fire-team leader after I made the rank of Corporal with the second platoon.

In April, Charlie Company was on very high ground when a US Air Force DC 4 dropped parachute flares and the front line was lit up like a giant football stadium. Very early in the morning we were awakened by enemy machinegun fire. I remember the haunting sound of an enemy bugler playing the popular tune, "Open the Door Richard."

During the night the friendly forces on our flanks had retreated and left Charlie Company unprotected and alone on the high ground. We soon regrouped, and escaped by marching down the hill through a village in flames. That seemed to be a prelude to the movie years later called, "Apocalypse Now." We marched for hours along a dark and deserted railroad track that eventually led us to a group of trucks that took us to an alternate line of defense.

Later in April, Charlie Company again went on the offense when the front line was not well defined. We crossed the line in front of the 7th Marines, and marched ahead on a five day patrol looking for the whereabouts of the Chinese and North Korean armies.

We moved along the Korean countryside about 20 miles ahead of our front and developed a perimeter on a hill that sat alone in the valley. Patrols were sent out in three directions and we never had contact with the enemy.

I remember, as if it were yesterday, hanging nine canteens on a tree branch and filling them from a stream at the bottom of our hill. A beautiful Korean lady and her Papa San stared at my every move.

It was May, and the front was very noisy with machine gun and artillery fire on both flanks, but it was quiet in our company sector. The 5th Marines always wore tan leggings and it was rumored that our enemy did not like to confront the troops with yellow legs. Anyway, we were soon relieved and moved down the hill to another sector.

While marching up through the valley, the sound of large green flies was deafening. Soon the obvious appeared as we saw bunches of abandoned dead soldiers from the 2nd Army Division Heavy Equipment. As we advanced up the hill, we passed by a tent full of dead Army Officers and other dead soldiers lying in shallow fox holes.

Later, two soldiers from this battalion came through our lines. They said the Colonel had told them to bug out, and they said that meant for everyone to do for themselves. We dug our fox holes at the top of the steep hill, and advanced forward the next day. Charlie

Company proved itself to be a formidable force in the battle on the Punch Bowl in June.

Both the first platoon under the direction of Pete Mc Closkey and the second platoon under the leadership of Chuck Daly led 'John Wayne' type charges against highly fortified enemy positions. They succeeded with minimum Marine casualties. My friend, Whitt Moreland, received the Medal of Honor for falling on a hand grenade to save the lives of other Marines on Hill 610. We had liberated South Korea before I was rotated home on the point system by September of 1951.

I am extremely proud to have served in Charlie Company during the Korean War. My service to our country in the United States Marine Corps gains in prominence as the years pass. The Marines that I served with were the most outstanding people that I have met in my lifetime.

"The Final Word" — A Blow-by-Blow Account

Korea, August 1951, Charlie Company, Fifth Marine Regiment. Charlie Company is back on the attack. We've marched all day. It is about 5:00 PM, and we are scattered along a trail that snakes around a large North Korean hill. We are in a hold position as we watch our Marine Corsair and Navy Hellcat pilots perform a ballet of bullets and bombs against the enemy force that dug in prior to our arrival.

These planes are a big help to soften the resistance before we make our ascent. They start by firing machine guns at the crest of the hill on enemy positions; then they launch the explosive rockets; next the 500 pound bomb is dropped; and the grand finale is the napalm that leaves a devastating liquid fire.

The planes have done their job and these brave and talented men have returned to their carriers or airstrips, in Japan, for a much deserved hot dinner and a warm shower.

It is 6:30 PM now, and we are saddling up to make our move. The second platoon is at the point of the attack and we will drop our packs and fix bayonets at the halfway point. The third platoon will handle our flanks, and the 1st platoon will act as reserve in the event we need help.

We never know the extent of enemy resistance until we reach the top of the hill, but we always succeed. Afterwards, we will dig our fox holes on the opposite side of the hill, and somehow find time to have a can of C rations.

Throughout the night we will have a 50% watch unless the enemy retaliates with a counter attack and then we will be back to a 100% watch. The enemy knows our position and it is likely that a few silent mortar rounds will greet us and disturb our sleep.

The Marine artillery will now be called in and they will fire shells all night in front of our position and on to the next line of resistance. Maybe we will have four or five hours of sleep. Tomorrow we will march on to our next objective

We take great pride in being able to live and sleep in a dirt hole, in preparing our individual meals, and in being able to survive the coldest of winters as well as the intense heat of summer. It gives one great confidence to be so self sufficient that you can live almost entirely off the earth all year long without cover.

We meet the enemy not far away, but close enough to see the whites of their eyes and smell the garlic from their rice ration. We are the dirty faced boy Marines of Charlie Company.

We carry a full pack with clean rifles, hand grenades attached to our cartridge belts, and extra bandoleers of ammunition are criss-crossed around our chests. Many of our shoes have holes in the soles from climbing mountains all year long, and our toe nails are black with the dirt from the mountain terrain.

We have great respect for the Marine and Navy pilots who helped us today, but feel we are the final word and consider our job to be supreme.

"The Marine rifleman looks up to no one!"

SGT PHIL. STREET, USMC
Jonesboro, IN

"Cook in Korea"

I SERVED ON ACTIVE DUTY FROM 1951 TO 1954 AND ARRIVED IN Korea in February 1952, assigned to the 7th Motor Transport Battalion as a cook.

On my first day in Korea I noticed we had no can opener. Big surprise! We fed almost 800 Marines at each meal because we had a company of Amphibious Ducks TAD from the Second Marine Division assigned to the First Marine Division.

Much of our food came in 16 ounce cans and we had to use a meat cleaver to open them. Bacon came in one pound cans. Both the top and bottom had to be removed and the side of the cans split in order to get the bacon out of the can.

Street needs a can opener

As a result, many of us had a lot of bad cuts from getting the bacon out of those little cans. When I left at the end of my tour 13 months later, we still had no can opener.

Our M-1939 Cook Ranges had to have 40 pounds of air pressure pumped into them to burn properly. We had no air compressor so we used a bicycle pump, which took forever.

The ranges used a generator (a piece of pipe) running down the center that needed to reach a certain temperature before the range would burn properly, with a blue flame. Otherwise, they just burned yellow and would flood, not heat right, and often catch on fire.

With outside temperatures as low as minus 40 degrees, the leather seal inside the bicycle pumps would freeze up. When it was zero weather, the leather seal was so stiff it could take 20 minutes or more, taking turns pumping, to get the 40 pounds of air in one range.

We had many ranges to pump up, so it took forever to get ready to light the burners. To make matters worse, when it was below zero, the generator was too cold to generate. We used a blow torch to heat the generator. It was often so cold the blow torch would not generate either. Then we had to burn gasoline in the little cup below the blow torch nozzle in order to thaw the torch.

Many times we had to keep repeating the process until the torch thawed out enough to work properly. We had one cook badly burned trying this. He took a funnel with gas in it with his finger over the end and just dripped the gas into the cup a little at a time.

When he repeated this the second time, the cup on the blow torch was too hot and ignited as soon as the gas hit it. He threw the funnel in the air and the gasoline came down on his arm and set his clothes on fire.

When we received water in the water buffalos (portable water trailers), it was frozen solid. So we had to climb on top and using a big steel bar, break the ice into chunks so we could melt it to make coffee and cook for the Marines.

The camp was blacked out at night except the mess tent which served the truck drivers who drove around the clock. We always had coffee and something to eat if they wanted it.

We had one cook who, for some reason known only to him, got mad if a truck driver would get coffee without asking him first. The cook had just bought a brand new, leather bound, short wave radio with AM on it from the PX truck that came around about once per month. It cost him at least a month's pay.

One night when he was on duty, the Marine who operated the camp generators at night came in and went to get some coffee without asking first. The cook chewed him out and would not give him any coffee.

The Marine left and about three minutes later our lights got real bright and all of them burned out. The cook's brand new radio was smoking and never did work again. A few minutes later the Marine came over and said something happened to the generator. He brought new bulbs with him and some flash lights for us to see to replace them.

He told me later, "That will teach him not to let me have coffee. I went back and ran the generator up to about 300 volts." I never told the cook what caused his radio to burn up. I figured he deserved it.

CPL ROBERT MCGUIRE, US ARMY
Daytona Beach, FL

Submitted by Rick Kennedy, Secretary,
Central Florida Chapter, First Marine Division Association

BOB'S 7TH GREAT GRANDFATHER, JOHN MCGUIRE, SERVED AS A scout for General George Rodgers Clark, and 14 of his uncles and grandfathers were quartered at Valley Forge with George Washington.

Nobody was going to keep Bob McGuire from serving his country in the great battles in Korea. Members of his family have been in every war in which the United States has been engaged, including Iraq where two grandsons have served and recently returned unscathed.

Bob McGuire was only 17 when he joined the United States Army on the 17th day of June in 1951. The Korean War was in full swing. It was appropriate for him to give up his title as President of the Junior Class at Tuscola High School in Tuscola, Illinois, to serve his country. After all, he was a descendant of a 7th generation U.S. Army Scout.

He served gallantly with the 32nd Infantry Regiment of the 7th United States Army Division in their great victories at Kumhwa, Iron Triangle, Old Baldy, and Pork Chop Hill in North Korea. His third day in Korea placed him in the assault on Triangle Hill, in Operation Showdown that was launched on October 14, 1952.

Only five men in Bob's assault platoon survived the onslaught of those 2,000 Chinese, after taking the objective and subsequently the largest artillery barrage of the war. His battalion was awarded the Distinguished Unit Citation, our country's highest Medal of Valor.

Later in May of 1953 on Old Baldy, Bob's Company Commander requested that he put together a rescue mission for 19 dead and wounded officers and enlisted men that the assault force could not reach.

Bob McGuire completed this mission under heavy enemy fire and was awarded the coveted Bronze Star Medal for bravery in playing a major role in bringing out the dead and wounded at great risk to his own life.

His Korean War experience gave Bob McGuire a feeling of pride and confidence that allowed him to advance successfully through civilian life using the same discipline and courage that he displayed in the great battles against our enemy on the mountains of the Iron Triangle.

His lovely wife Nancy has treasured his love and loyalty for 54 years. Together they have six wonderful children, 17 grandchildren, and 18 great grandchildren. They have resided in Florida for over 40 years, where Bob has been a very successful business owner in the pest control business.

Left: McGuire in Korea in 1953 and (right) in 2008

CPT JOHN A. PALESE, MD, U.S. ARMY
Milwaukee, WI

"A One Year Old Child"

I WAS THE BATTALION SURGEON FOR THE 3RD BATTALION, 65TH Infantry Regiment of the 3rd Infantry Division from Sept. 1950 to Oct. 1951. This was an all Puerto Rican Regiment with the exception of some of the officers, like me, from the States.

We started in Pusan and fought our way up into North Korea. The incident I am about to relate took place during this period. I cannot recall dates or sites. As you know, the infantry is on the move on a daily basis, depending on combat circumstances, and we move with them.

We had just moved into our assigned position when a Korean man and his wife, carrying a one year old child came into my aid station seeking medical attention for their son. With the assistance of a Korean interpreter I learned the child had been ill for four to five days with a high fever, a severe cough, and was now having trouble breathing.

Upon completion of my examination, I diagnosed it as acute Pneumonia. My problem was how to treat a one year old, because most of our medical

Palese taking wounded to MASH by helicopter

supplies were for adults. Our main source of antibiotics up to then was the Sulfa drug family. Fortunately, the Army had just been provided with Penicillin that was to be used for military personnel only.

Forgetting the rules, I injected a good dose of Penicillin into the boy and they departed. As we moved around and days passed, I wondered what happened to the boy. I often wondered how they found us in the first place. I was surprised when two weeks later they showed up in my aid station again.

We had moved several times since our first meeting and they had to walk miles to get to us. The little boy was fully recovered and they were there to give me a large sack of chestnuts for treating their son. How they found us both times will remain a mystery to me forever. All's well that ends well.

> "They that can give up essential liberty to obtain a little temporary safety deserve neither liberty nor safety."
>
> —Ben Franklin

PM2C FRED FURUICHI, USN
Fairfield, CA

"The Best Four Years of My Life"

FROM THE TIME I WAS A LITTLE KID, I WAS DESTINED TO BE IN the Navy. I don't know who put me in a sailor uniform for the family picture.

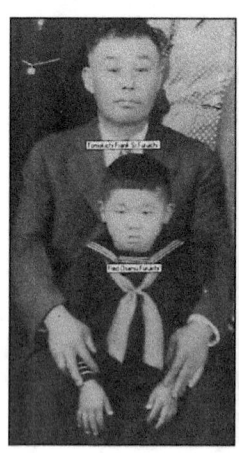

That's my paternal grandfather behind the little guy standing in a navy outfit. He was a farmer in Japan and thus he continued to seek his fortune in the early 1900's by farming in Los Altos, Santa Clara County, California. Grandfather was a good man, a hard worker, a Founder of the Buddhist Temple, and a leader of the community.

In 1948, at age 19, it was my turn to serve my country. I decided to volunteer for the Navy to avoid being drafted into the Army. I was sent to San Diego for Boot Camp and then to Aviation Service

School in Memphis, TN. At Memphis I qualified for Aerial Photography School in Pensacola, FL.

In 1950 the Communist North Korean Army invaded the Republic of Korea (ROK). North Korea declared War on South Korea and marched south of the 38th parallel which was the North/South Korean border. They swarmed into South Korea with little or no resistance from the ROK Army.

Within a month's time, I received word that the US Navy was forming an elite "Combat Camera Group" and invited Navy photographers to volunteer for the Combat Zone to photograph the Navy's part of the conflict. My classmates from Naval Aviation Photo School were thinking of joining the group and kept urging me to go with them.

At first, I was reluctant, but it was an opportunity for me to get sea duty, a coveted duty for which every sailor lives and dies. The Navy needed 30 volunteers for this group made up of still photographers, yeomen, movie cameramen, production, and newsreel photographers. Going overseas as a Navy newsreel photographer was very appealing, so I finally volunteered to join my classmates.

We were the "Greenhorns" of the group. All of us came together well because we were mostly in our early 20's or late-teens and our commander was barely 32 years old.

Getting our necessary photo equipment along with special flight skins attached to our dog-tags and other gear took about two weeks. This gave us a chance to learn hand-to-hand combat from the veteran Marines on the base.

The first contingent of our group left for the Far East soon after our gear arrived. A week later the rest of us boarded another Navy cargo plane. It carried our photo gear and huge crates of supplies for our Marines. There was a constant roar from the propellers during every minute of the flight across the Pacific Ocean.

General MacArthur's battle plan called for an amphibious invasion of the Inchon Harbor on the west coast. It was a brilliant move that slowed the communist assault and nearly cut-off 40,000 to 50,000 enemy troops in their march to the South.

My combat camera group missed the Inchon Harbor Invasion. When our small unit of three correspondents finally arrived in Korea, we saw the aftermath of the successful US Navy's amphibious landing.

The landing area was strewn with debris, abandoned vehicles, and bombed buildings. The landscape had huge holes made by the 16" projectiles fired off-shore by the battleship USS Wisconsin. The holes were large

Furuichi and friends aboard the battleship USS Wisconsin.

enough to swallow an entire house. The enemy didn't have a chance against the Navy's artillery. No bodies or enemy soldiers were in sight.

The North Korean communists were actually in rapid retreat. Securing the beach-head made it possible for our supply ships to unload cargo. This opened the port for more troops to land and thus secured Inchon Harbor as our entry port for supplies and fuel for our armored tanks, trucks and jeeps.

The massive assault also made it possible to advance through the capitol city of Seoul and beyond. While I was passing through, I saw that most of the city had been leveled earlier by the North Korean Army which had destroyed everything in its path. It was an unbelievable sight.

By the time we caught up with the US Marines, they were holding their positions at the spear-head of operation. At the front line, the Marines were flanked by the Turkish Brigade and the Ethiopian Army. We got to know the Turks, Marines, and the Ethiopians really well.

We shot hours of footage while aboard Aircraft Carriers, Battleships, Cruisers, Destroyers, and even Mine Sweepers. There was always a team of photographers covering the Peace Talks at Panmunjom. The most exciting assignment for me was covering the takeoffs and landings on the flight deck of the USS Philippine Sea, and going on bombing runs off the carriers.

I flew many bombing sorties while riding in the Gunner's seat on an AD Skyraider aircraft. The carrier would turn into the wind while the plane handlers would position us for launch on the flight deck and hook our aircraft to the catapult cable. The navy aviator would "rev-up the props" and in an instant the bomber would be "flung-out-into-space." It was a thrill to be shot-out like that and to be flying at 80 knots in a few seconds of time.

As we approached for landing, we could see the carrier turning again into the wind. Fear set in when I could only see a tiny flat-top carrier. It looked to me like a postage-stamp, floating on the sea. It seemed impossible for the bomber to safely land on a postage-stamp. I quickly looked for the "eject button" but there wasn't any such thing.

Gritting my teeth, I braced myself for a crash landing. I heard the flaps go down (quite noisy) and the "tail hook" dragging. The pilot made minute adjustments.... Silence.... and abruptly the plane slammed onto the wooden deck of the Carrier — our forward motion suddenly arrested by the hydraulic cables. Wheeeeeee — a safe landing!

We were nine to ten months into our campaign when my Commanding Officer, Lt. Commander Gould, summoned me to his office in Tokyo. Expecting the worst, as I reported, my first thought was, "Where did I screw-up?" Thankfully, it turned out to be a questioning session and not a reprimand.

He wanted to know about my possible relationship with the Japanese Nationals. His men had told him that I had some Japanese relatives nearby who survived WWII. He asked me where in Japan they lived. After a few more questions, he ordered me on a special mission for 30 days.

The order read "The Mission: To locate and examine the vanquished WWII Japanese Imperial Naval Facilities of Nagoya, Japan." I was to take no "official" pictures nor write a report. I was to just visually inspect the condemned Naval Yard for my own edification.

Left: Picture of Fred, center, and two Turkish Brigade soldiers; Right: Fred on flight deck of The Philippine Sea

Basically, it was a bogus order. It was just a way for me to visit my grandparents, uncles, aunties, and cousins on the Navy's Order. I indeed welcomed the orders.

First, I let my families know about my mission. Then I shopped at the US Army Post Exchange and filled my sea bags full of American goods such as cans of ground coffee, pounds of American chocolates, cartons of American cigarettes, and the like.

I put on my white sailor uniform and white sailor hat, of course, and boarded a train from Tokyo to Nagoya.

At the Nagoya station I met my maternal grandmother, Kiyoko, for the first time, as well as my auntie and her husband, and a bunch of cousins whom I had not seen since their parents repatriated to Japan from California about two years before WWII broke out.

I had a grand time for 30 days getting re-acquainted with my cousins, uncles, and aunties while living with my maternal grandfather and grandmother. I was truly grateful for the 'Navy Order' and that once-in-a-lifetime experience.

Commander Charles Gould had great compassion to do something so generous for someone in our group. All of us at CCG knew he was a self-made man. He was raised in an orphanage, foster-homed, self-taught, and put himself through college.

Upon my return to duty he said he had never known his biological parents or siblings (if any) and had never met his grandparents, uncles or aunties. He never knew who they were. My epiphany: CMDR. Gould COULD NEVER have a once-in-a-lifetime opportunity like mine, but he gave that opportunity to me.

Cdr. Gould was happily married with two children and lived in New York City. After he served in Korea as a Naval Reserve Officer, he returned to civilian life and rose through the ranks of the newspaper business to become Editor-in-Chief of the New York Herald and the San Francisco Examiner newspapers.

Fred with relatives at Nagoya Station

During my life of 79 years, serving in the US Navy was by far the best four years of my life for the following reasons:

First of all, I was able to serve my country in the US Navy which I have loved since I was a little boy. "Join the Navy and See the World" the slogan said. Indeed, they kept their promise.

While overseas and separated by a vast ocean, I really got homesick for the first time. It was a realization of how much my Mother had loved me all my life and how deeply she feared for my life while I served in the Navy. When I left home she begged me to return to her and in one piece. Thankfully, I was able to keep that promise.

In combat I never doubted that my buddies would "get-me-out." Because I was in the company of MEN of HONOR.

The Navy experience reinforced my parents' examples of honesty, loyalty, selflessness, compassion, integrity, love, and more.

Thanks to the compassion of Lt. Cdr. Charles Gould, I was the only sibling of five children to have met and lived with our maternal grandparents, who never left Japan.

President Truman extended the G.I. Bill of Rights to Veterans of the Korean War. Without the G.I. Bill, I would not have gone to college. Moreover, without the Cal-Vet Bill, I would not have earned my Masters Degree from Harvard University. I am a retired Landscape Architect and the following I truly believe:

> "One must seize the opportunities as they appear! Each, will enhance one's life."
>
> —Fred Furuichi,
> Photographers Mate Second Class

SGT ANTHONY P. RABASCO, US ARMY
Scarsdale, NY

THIS IS THE SEVENTH YEAR I HAVE SPOKEN TO THE CADETS AT West Point Military Academy, West Point, New York. Every year they send a letter requesting that I come back.

There are about 150 Korean Veterans who attend. They call you by name and assign you to a class of 15 Cadets. You speak to the Cadets for one hour about your experience in combat. The cadets are very interested in hearing about war from a foot soldier.

I arrived in Korea in 1951 with the 101st Battalion. After making the Inchon landing, I was detached from my unit (TDY) and assigned to the 24th Division supplying Antrack radio. We moved through Seoul up to the north central part of Korea. My unit relieved the 561st Signal, which was attached to the 1st Cavalry Division.

During the winter months of 1951, I supported communications for the 1st Marine Division on the Han River. Moving forward, I supported communications with the 19th Brigade for about two weeks.

Rabasco — It's a lonely life on the mountain top

Later I was assigned to the 6th ROK Division of South Korea, along with the English Army troops called Newcastle.

The Chinese broke through the front line and we retreated on foot, moving 60 miles south to Chuncheon in two and a half days. We regrouped and went forward with the 7th Division to the 38th Parallel, the dividing line between North and South Korea.

It was on that occasion that we were lucky enough to have a squad of infantrymen who gave us some support and helped us avoid being ambushed.

We were on the highest mountain ranges where it was easier to transmit and receive radio signals, but getting supplies was difficult. It was important to keep our communications open since we were the only link between the soldiers in the field and headquarters.

Each radio team had only three men assigned, so we were basically alone. Being alone in a hostile environment can bring all sorts of fears to your mind. That was hard to deal with. Also, we had to carry all our belongings on our backs. That included a tent, sleeping bag, water, a weapon and a generator for the communication radios, which weighed about 55 lbs.

Rabasco — A place to call home and a haircut

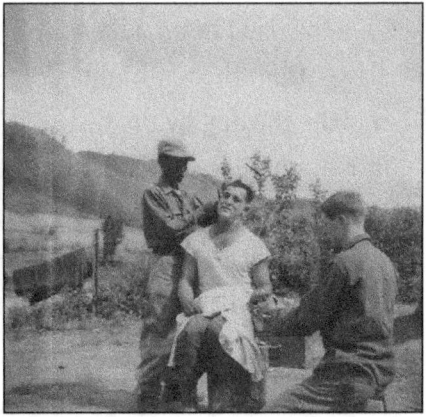

It was very difficult to get supplies and drinking water to us. Five gallons of drinking water had to last one week among three soldiers. We washed with a washcloth because we could rarely get to a shower in a camp. If I took a hot shower five times within the course of a year, that was a lot.

A change of fresh clean clothes was difficult. Seldom did we have the opportunity to get a warm meal, unless we were close to a unit that we were supporting. I relied mostly on field rations to survive.

July and August in Korea are the months when the monsoons came. Continuous torrential rains flooded the land and even penetrated our canvas tents. It was like someone spitting at you.

We would dry out our socks by putting them between the waistbands of our pants, hoping our body heat would dry them out. The ground was so wet that we had to put our rain ponchos underneath our sleeping bags so we could keep somewhat dry.

As we moved through the villages, going north, the Korean people abandoned their homes. You would notice the soil was packed down along the road next to their huts. Then you would see crevasses in the soil. Digging up the soil, you would find huge crock-pots buried beneath the earth.

One day I heard a dog screaming. I looked through my field glasses from the top of the mountain, and saw that they were choking the dog by the throat. When I arrived at the foot of the mountain, they started burning the dog on the fire and sapping the fur off the skin with a stick. They boiled the dog in a huge can with all his internals, and then they ate the dog.

I arrived back in the United States on April 3, 1952.

A3C ROBERT RABER, USAF
Sequim, WA

Senior Cook

I JOINED THE USAF ON A FRIEND'S RECOMMENDATION, JUST prior to my 18th birthday. I thought I had things tough at home on the farm. Boy was I wrong!

I was sent to Lackland Air Force Base, in Texas, for Basic Training and was asked if I had a high school diploma. They were looking for pilot trainees. I didn't have one so they sent me to Food Service School in downtown San Antonio, TX.

After that I was sent home for 30 days to Port Angeles, WA. Then I was told to report to Camp Stoneman, CA, to be sent to the Far East. My Dad had told me I was born in Wisconsin which was "back East" so I thought that was where I was going; But why go by ship?

Anyway, we landed in Yokohama, Japan and my APO number was different from my buddy's. I asked why and the Sergeant on the dock said I must have been 18 when the orders were cut. Since my buddy was younger than me, he got to stay in Japan for 36 months, as far as I know. I didn't see him again.

Raber — Cook & Piano Player

Boy was I surprised to be in Japan rather than Wisconsin! Since I had turned 18 when I hit Boot Camp, I was being sent to Korea for 12 months. Upon arrival in Korea, I was assigned to the 67th Food Service Squadron as part of the 67th Tac Recon wing at Kimpo AFB K-14 above Seoul, South Korea.

I remember an event that happened one evening when "Bed Check Charlie" flew over the Air Field and a "Red Alert" was sounded. Several times a week at 10 P.M., or Lights Out, this Commie Piper Cub would come over and fly up and down the air strip. He would throw small bombs from the cockpit to pock mark our air strip.

We could see him quite plainly in the moon light. The machine gun emplacements around the base were not allowed to fire on him until he got off the base, so no personnel would be hurt if his plane crashed on base.

We had a fellow in our Squadron who was from Georgia and, of course, his nickname was, "Georgia." He was in a trench with our Squadron when they opened up the 50's. Nobody was hitting him, so Georgia got out of the trench, ran over to the gun emplacement, and asked to borrow an M-1 rifle. They gave him one, and he took aim. In that instant the guy on the 50's said sarcastically, "I suppose you think you can hit him." Georgia touched one off and must have hit the fuel tank. The plane exploded and the rest of us applauded.

He turned to us and said, "Where I come from if you don't hit what you're aiming at, you don't eat." Needless to say, we all had a bit more respect for Georgia from that evening forward.

While on duty one afternoon, a fellow in a different uniform came into the dining hall looking kind of lost. Nobody paid any attention to him, so I approached him and noticed an Australian Air

Force patch on his shoulder. He politely asked if he could buy some bread from us as he had received some "Sweet Meats" from home, but he couldn't find bread anywhere. I told him he didn't have to buy it because we had a squadron bakery right behind the dining hall and they made our bread and pastries fresh every evening for the next day's servings.

I asked, "How do I find you?" He told me where the Aussie Squadron was located and said to ask for, "Casey." I sealed six loaves of bread, made fresh from the oven that evening, into a Kraft wrapper and headed out.

I found him by way of another Aussie who I believed to be an officer. When we found Casey, everybody snapped to attention. The officer turned out to be a two star general in the Australian Air Force.

After sampling some, "Sweet Meats," Casey asked if I'd like a "Brown Bubbly," their name for beer. We went to a larger tent that had a wood floor, and upon entering the tent I heard people singing around a piano.

I asked Casey where their piano player was and he said, "We don't have one. Do you know how to play the piano?" I said, "Yes," so Casey yelled, "Hey you blokes, this cook plays the piano." So I sat down and played, "Now Is the Hour" over and over and over, all evening long.

It was the only tune they wanted to hear, as half of the squadron was rotating home the next day. After that, the Aussie Recreation Room was my second home.

This was the winter of 1952. The only foul weather gear I had was a fatigue jacket. I mentioned that I had guard duty that evening, so they willingly lent me a flight suit and boots. They were wonderful friends. They said they didn't want their piano player to get frost bite. It turned out to be a very pleasant experience in a War Zone. I was at Kimpo from June 1952 to June 1953.

    Back home, I worked in the local pulp mills, in Port Angeles, WA. Later on, I ended up working road construction driving a dump truck, and running a loader for more money, as my family grew. I also worked for Clallam County Road Department and drove a log truck when no other work was available.

INTRODUCTION

THE FOLLOWING SEVEN STORIES* WERE RELATED FROM VIDEO records of a panel discussion hosted by Clayton State University in Morrow, GA on October 2, 2003. The panel included six Korean War veterans who served in the Chosin Campaign.

The discussion was held in honor of the late Marine Corps General Raymond G. Davis Sr., holder of the Medal of Honor for his actions during the Chosin Campaign. Members of Davis' family were also present. The beginning of the panel discussion started with excerpts from a video taped speech given by General Davis prior to his death.

Left: Veteran's panel;   Right: Jackson and General Davis

---

* The seven have shaded boxes around the names of the soldiers.

GENERAL RAYMOND A. DAVIS, SR., USMC
(Deceased)

*As a Lt. Colonel, Ray Davis received the
Congressional Medal of Honor for actions during the Chosin Campaign.*

I WAS ZIPPED UP INTO MY SLEEPING BAG ONE NIGHT WHEN I noticed a marine standing up against the silhouette of the sky. I sat up to tell him to get down and a bullet came through my sleeping bag and creased my skull — so I got back down! But the Lord was taking care of me.

I had six close calls happen to me. One round through my knee, one round across my skull, a mortar shell knocked my helmet off, a bullet went through my helicopter, and one went through my side. But, I never missed a day of work!

So the Lord was taking care of me and I think probably led me when I retired. I taught Sunday school for 20 years.

With regard to the White House ceremony for the presentation of the Medal of Honor, General Ray Davis stated: "There were three of us meeting with President Truman. He said to one of the guys, 'I'd rather have this medal than be President.' But we didn't trade with him!"

SSGT ANDREW B. JACKSON, USMC
Fredericksburg, VA

I HAVE TWO THINGS TO TALK ABOUT. THE FIRST WAS SOMEWHAT comical, but at the time we did it we didn't know if it was so comical or not. After we got to the Chosin reservoir in North Korea, we discovered we needed more supplies from back at Wonson Beach. A motorcade was formed to go back down to Wonson to pick up supplies and ammunition and bring it back up to the reservoir.

When we arrived and got the trucks loaded, we received word that the Chinese had completely surrounded the reservoir and that there would be no way we could get back. We knew that they were running short of supplies up there, especially ammunition.

The Greeks were flying their boxcars and we talked them into flying us up there. But after we were airborne, we found out there was no way to land those planes at the reservoir. It just happened that Marguerite Higgins (Maggie; the war correspondent) got onto the boxcar that I was on. Some of you may have heard of her.

We had a load of ammunition on our boxcar. The pilot asked us what were we going to do and I said, "I don't know but we're going to get out of this plane somehow!" That's when he said, "Ok, we'll get you out."

He opened the back door of that boxcar and flew right at ground level. When he came to a hill, or a little mound, he took that plane straight up! All the passengers, along with the ammunition slid out the back of that plane.

The snow was about six or seven feet deep, and when we hit that snow it was like a great big sheet of styrofoam as we slid down that mound. We landed right in the middle of the Chosin Reservoir. So, that's how we got the ammunition up there!

The second thing happened when I landed at Wam-a-du Island on September 15th, 1950. When we fought, my particular job was communications chief, and also FO. For those who don't know what FO means, that is a Forward Observer. FO's observe the front line and call in the ground artillery, the air strikes, and the gun fire from the big guns out on the ships.

As we worked our way up toward Chosin, Lt. Col. Ray Davis needed a FO because he had heard there were two companies of Marines trapped, and we were going to go in there and get them out.

The Chinese were pretty good at tapping our communications lines. When we got up to the reservoir, Lt. Col. Ray Davis and I figured out a little code. I know all of you have heard of 1 Indian, 2 Indians, 3 Indians, and so forth. Well we did that. If Ray told me he needed 2 Indians that meant he needed the gunfire 2 clicks up. If he needed 2 squaws, that was two shots down. If he needed a TeePee, that was to the left, and if he needed a papoose that was to the right.

I was walking behind that big tank up there giving the direction of gunfire. I was dropping those rounds right in front of Lt. Col. Ray Davis and his troops as he spearheaded his troops through the Chinese.

That was my job as FO in Korea with Lt. Col. Ray Davis.

[Editor's Note: Staff Sgt. Andrew B. Jackson, USMC was also a POW at Chosin Reservoir and was rescued after two weeks of torture.]

MAJ ROLAND A. MARBAUGH, USMC
Conyers, GA

I NOTICE GENERAL DAVIS SAID EVERYTHING WAS FROZEN. IF YOU think you don't get thirsty in cold weather, think again, because you do. And eating snow didn't seem to be the right thing to do. There was always 'weed seed' in it and it just wasn't satisfactory because you couldn't dig a piece out of your canteen.

One time we found a dug well around dusk, but couldn't see down into it. I hooked up my canteen cup and got a cup of water. We were debating whether to drink it or not when the battalion doctor, Dr. Moon, walked up to us. He licked his lips and said, "I proclaim this water fit to drink," and drank it!

Then along came a guy with a flashlight and when he shined it into the well there were two dead Chinese down there about that big around (huge gesture). We never told Dr. Moon about the dead Chinese.

Another thing comes to mind as I see the Sergeant here with us. I once heard General Davis once say if you could run a squad as a Sergeant you could run a division. On the way down the mountain, I was the Headquarters Commandant of the 3rd Battalion, 5th Marines and I had one heck of a job.

Many of the Marines in Korea had never been in cold weather before. I was one of the fortunate ones, because I had been in cold weather all my life. When I come upon two of my men sunk down in the snow, I jerked them up, gave them a slap across the face, and a boot in the rear end saying, "Now you keep moving or next time

Marbaugh at Camp Pendleton, CA; Waiting for plane to Korea — 1950

I'm going to give you more than a boot."

One says "Ok Sarge, Ok Sarge" and the other one says "He's a Captain." And the first one says, "Gee I'm sorry Sir." I said, "Forget about it kid, you just paid me a great honor!"

LT WILLIAM HALL, USMC
Atlanta, GA

WHEN THE NORTH KOREANS INVADED SOUTH KOREA ON THE 25th of June, 1950, I was a carrier pilot assigned to MCAS (Marine Corps Air Station), El Toro. On 3 July we were alerted that we were going to Korea.

Due to the annual transfer and a shortage of people, we had a commitment for 24 Corsairs, and 30 pilots, plus two LSOs. I was one of the paddle waivers an LSO (Landing Signal Officer).

We departed San Diego on the 13th of July aboard a Jeep Carrier which is 500 feet from the 'pointed end' to the 'round end,' as we used to tell the Navy. My first combat mission was on the 6th of August supporting the First Brigade, which were all the folks the First Division had in it at the time.

They couldn't call it a Division because there weren't enough folks, so they called it a Brigade. We were at the Pusan perimeter and I was in charge of the first wave that landed at Blue Beach at Inchon on 15 September. Then we moved around to the eastern coast to support the Marines at the Chosin Reservoir area.

We had 24 Corsairs aboard, and 24 on another carrier just like ours. So we had a two carrier task force of Marines doing close air support. We also had a couple of airplanes ten miles toward Vladivostok flying at ten thousand feet in case anybody came down.

In supporting the first division and supporting the Army troops also, we launched at pre-dawn, six aircraft from each squadron so we would fly over the First Marine Division by daylight.

On one of these occasions an Air Force Transport pilot came in and got there before the Corsairs arrived. He could see the Chinese shooting into an air strip that the Marines had built. (Building the air strip was contrary to what the Army Corps told General O. P. Smith to do, but it saved a lot of lives both Army and Marines.) The pilot said that as soon as he flew over the First Marine Division the Corsairs came in, and the Chinese disappeared.

We had Corsairs, and the Navy had Corsairs and ADs. They also had some jets, but the ADs and the Corsairs were the main close air support weapons out there. We relieved each other all day long until dark. We did not have night flying because the carrier was not rigged for it.

We carried 800 rounds of 20 mm cannon, 8-five inch rockets, and either a 1,000 pound bomb or a 150 gallon tank of napalm on the left pylon. To make sure we could stay up for a long time, we carried a 150 gallon tank of gasoline on the right pylon. So, with 233 gallons of internal fuel plus 150 external, our average flight was a little over three hours.

Only two pilots got stuck out and didn't get back in time to fly the 35 to 40 miles out to sea. Neither one of them made the first pass and aborted. I know because I was one of the pilots that made it. I was given a DFC (Distinguished Flying Cross). I found out about it six months ago (April, 2003). I didn't even know I'd earned it for that flight. Later, for making a night carrier landing when the carrier wasn't rigged for night flying, they gave me another DFC.

We had a lot of activity up there. I think I dropped a napalm tank on a Chinese platoon standing in formation, and whether they were going to do something or whether they had frozen, I didn't know. Anyway, they got warmed up pretty well, I think, with 150 gallons of napalm.

MAJ NORMAN A. SABEL, US ARMY
Stone Mountain, GA

I AGREE, IT WAS COLD UP AT THE CHOSIN RESERVOIR IN KOREA. I had trench foot during WWII at 'The Bulge' battle in Europe, and in Korea I had frozen feet. It didn't bother me because I knew what it was.

I was originally in Japan and I had one of the largest radio and telephone repair shops in Japan. I had memorized over 20 thousand parts of signal equipment. Although I was Army, I made the Inchon invasion and went into Wom-bec Island with the Marines carrying messages. When we got to Inchon, we landed and went to Hang-yang-nee.

My company commander called me in and said, "We want you to go back to Inchon and pick up our trucks and then lead the convoy back." I argued with him and said, "I'm not a truck driver sir." Then he said, "You are now!" So I drove a truck during the complete Korean War. I was transporting wounded and sometimes coming back fully loaded with trailers, hauling ammunition, food, anything you could think of, and repairing tires, too!

A sobering thing happened as I traveled through South Korea.

Sabel: Left: Trench foot in WWII; Right: Frozen feet in Korean War

When I first took over the truck, I was sent to Seoul. There were pill boxes in the street and fighting going on up and down and across the street. I pulled into an alley where I was supposed to go and stopped.

I walked past a sleeping guard and through the back door of a warehouse. When I talked to the Korean Colonel, he asked, "How you get in here?" I said, "I walked through the back door." He said, "What about the guard?" and I said, "Well he's out there." He went outside and saw the guard asleep with his rifle lying across his lap. This Colonel took out his gun and blew the guard's brains out right there and then.

You can believe I never went to sleep on guard duty!

I went up to the Yalu River near China on a run, and came back down. We were to the right of the Marines. I helped evacuate them down to Hungnam where the Navy saved us. They picked us up and took us out of there.

CPL DEWEY L. NORTON, USMC
(Deceased)

MY NAME IS DEWEY NORTON AND I GREW UP IN SAVANNAH, Georgia. It was there that I joined the United States Marine Corps Reserve while I was still in high school. I came into the Marines through the reserves.

When we were all activated in the latter part of July 1950, I found myself in Korea not knowing what I was going to do. I was trained to be an infantryman, and re-trained to be a 60mm mortar-man.

When I got to Hagaru-ri with the First Marines, the First Sergeant told me that I was going to be a Chaplain's assistant. When I said, "What does a Chaplain's assistant do?" He replied, "Your primary duty is to be the Chaplin's body guard, and if anything happens to him, you'll come up before Captain's Mast." I took a swallow and I said, "Well, I'll give it a try," I not only was the Chaplin's body guard, I was also his flunky to do almost anything he wanted done; to help out in any way.

Norton (left) and friends: Chaplin's Assistants, 1951

There were 26 chaplains who served the First Marine Division at that particular time. Seven of them were in the Yudam-ni Valley at one time. They each had a congregation of about 1,000

Marines because there were about 25,000 marines in the First Marine Division at that particular time.

The Chaplains were not immune to enemy activity if they had the courage to wear a helmet. We had three Chaplains at the Chosin reservoir that were wounded and one chaplain's assistant was killed, as well as three other assistants who were wounded.

Those of us assigned to the Chaplains did things other than just be their body guards. We manned road blocks and were set up in the valleys in case the enemy troops came through at night. We climbed the hills and brought down dead and wounded soldiers. If a battle happened to be going at the same time we were on top of the hill, we helped whatever unit was up there in the fray.

Hill 1282 was my first experience with combat. A jeep driver from Able Company had to go up to find out where two of his platoons were but he didn't want to go up in the middle of the night. On the 27th of November, the first night the Chinese hit us, he asked, "Chaplain, will you go with me?"

He was an Indian boy from Montana and his name was Johnson. I thought how strange it is to have an Indian named Johnson. We went up the hill together and fought with the last 12 men of Easy Company that morning. While we were up there, the thing that saved our necks was two Corsair pilots who flew over the hill, saw what was happening, and took out the enemy for us.

We helped move some of the wounded down and then helped re-supply those two platoons that were still there on Hill 1282.

SGT ALVIN E. LANDERS, USMC
Warne Robbins, GA

TO START FROM THE BEGINNING, I WAS STATIONED IN PANAMA and then I was transferred back to the States. We came into the Brooklyn Navy Yard at about 10:00 o'clock. They sent us to the barracks and notified us about 4:40 — 5:00 o'clock that "X" number of us were going to Lejeune.

I was going to be sent to Camp Lejeune, SC. Two days later we boarded a train and went to Camp Joseph H. Pendleton in California. While going across country, our Lieutenant told our Sergeant, "Let Landers get off the train anywhere he wants, to get a beer or whatever, just as long as you really watch him when we go through Texas."

He said that because I hadn't been home since the latter part of 1947 and it was now 1950! When we stopped somewhere in Texas, I jumped off the train. It just happened that the Lieutenant was standing there, and said, "Where are you going?" I said, "I'm going over there and get me some cigars," and he said "Ok." When I came back wagging a case of Jax beer, he said, "Where are your cigars?" I told him they were sold out.

We continued on to Camp Pendleton and stayed there for two or three weeks. I was in the 4.2 mortar company, 7th Marines. While there, we got our mortars and spent about three days learning how to set them up and how to fire them. Then we loaded them aboard ship and headed for Korea.

We didn't make it to Inchon, Korea, because we hit a typhoon that caused a lot of damage aboard our ship. We had some casualties too. Instead, we went to Kobe, Japan, where the Japanese unloaded the cargo holds.

While unloading, our Company Tech Sergeant looked surprised. Then he ran and booted this one Japanese, who went down and spread-eagled himself. When he jumped up they hugged like long-lost brothers.

Come to find out, our Tech Sergeant had made the Bataan Death March, and this Japanese was his prison guard when he was in the POW camp there in Kobe during WWII.

CPL FAUST SOFO, US ARMY
Staten Island, NY

"Pork Chop Hill"

I DON'T KNOW IF MY STORY HAS ANY MERIT OR WHAT YOU ARE looking for but this is a tidbit of my story. When I was leaving for Korea, my wife (who was my girlfriend at the time) gave me a Saint Christopher Medal to protect me from harm.

I was with the 7th Infantry Div, Co "G", 17th Infantry Regiment (White Buffaloes). On or about early April 1953, we were sent up "Pork Chop Hill." When we got close to the hilltop, we started to get bombarded. We all dove to the ground as mortars were exploding all around us. Later on that day, I notice my Saint Christopher Medal was gone. It had fallen from the neck chain holding my dog tags.

Then about a week later, we were rushed back to "Pork Chop Hill" and were told to hold it at all cost. Again we were bombarded. But while diving to the ground something shiny caught my eye. I couldn't believe it, but there it lay! It was my Saint Christopher Medal. I was blessed.

On this day, April 17, 1953 (Easter Sunday back home), one of the biggest battles of the Korean War took place on "Pork Chop Hill." Many companies lost hundreds of men. I was one of the lucky ones to make it back down that hill.

Actually this hill became famous. "Pork Chop Hill" was an important strategic hill at that time and we were told to hold it. Later on the movie called "Pork Chop Hill," starring Gregory Peck, was made of this battle. The fighting went on and on. We kept hearing that an "Armistice" was being signed in Panmunjom. The battle continued through April, May, and June.

Then July 27, 1953, on my birthday, Korea got quiet. This was one of the best birthday presents I ever had. Thank God the Armistice was signed that day, and my Saint Christopher Medal was safely back with me.

CPL. Pat Finn and CPL. Faust Sofo — buddies in Korea, 1953

CPL BOB SWEENEY, USMC
Waycross, GA

MY OUTFIT WAS I CO. 3RD BN. 7TH MARINES. I SERVED IN THE SECOND squad as a scout, then made first squad leader. The 1st platoon and 3rd platoon were stationed on the outpost. The 2nd platoon was on the MLR and went around the OP (outer perimeter) every night on patrol.

We covered the front lines from a hill called the HOOK all the way over to outpost #1 next to Panmunjom. Some of the hill's names I cannot remember.

The poems on the following pages were written by two of our men.

"The Night Before Christmas" and "Western Korea" were written by Scochie. That is what we called him over there. I did not know his name until Higgins sent me the poems. His real name is Bob Darling and his ashes are buried at Arlington Cemetery.

"Outpost Dagmar" was written by Frank Youngerman on February 22, 1953. On the outpost they were warned about the enemy attack by the second platoon that was on patrol in front of their lines.

The North Koreans got a surprise by having to fight on two fronts. They did not take the hill. The Marines that I served with will live forever in my mind and heart. This is the best I can do for you. It is the other men that everyone should remember.

Semper Fi,
Bob Sweeney

THE NIGHT BEFORE CHRISTMAS

Written by Cpl Bob Darling, USMC
Item Co. — 3rd Battalion — 7th Marines
1st Marine Division — Western Korea
December, 1952

'Twas the nite before Christmas and up on the hill
The lines were quiet the goonies were still

The shoe pacs were hung in the bunker with care
In hopes that they'd issue each man a new pair

The weary Marines were sacked out in their beds
With visions of girl friends dancing in their heads

When up on the ridgeline there arose such a clatter
A Chinese machine gun had started to chatter

I rushed to my B.A.R. and threw back the bolt
The rest of my buddies arose with a jolt

Outside we could hear our platoon Sgt. Kelley
A hard little man with a little pot belly

Come Perkins, come Sweeney; come Joe and Watson,
Up Higgins, up Maxwell, up Bailey and Dobson

We stumbled outside in a swirl of confusion
So cold that each man could have used a transfusion

Get up on that hilltop and silence that Red
And don't you come back until you're sure that he's dead

And putting his thumb up in front of his nose
Sgt. Kelley took leave of his shivering Joes

But we all heard him say in a voice soft & light
"Merry Christmas to all and may you live thru the night."

WESTERN KOREA

By Cpl Bob Darling, USMC
Item Co, 3rd Battalion — 7th Marines
1st Marine Division — Western Korea
December, 1952

Far across the frozen ditches,
there's a land we know so well.
Where the Marines of the 1st Division,
gave the Chinese Commies hell.

Frozen hands and frozen feet,
frozen rifles and frozen meat.
Rotten teeth and diarrhea,
We got them all in Western Korea.

Rubber boots and fur lined parkas,
Frost-bitten and frozen carcass.
Warm your beans with a Ronson Lighter,
That's what's known as a combat fighter.

General Pollack is our CO,
in this land of ice and snow.
The coldest spot in God's creation,
We're freezing here for the United Nations.

Now you've heard our gripes and bitches,
in this land of frozen ditches.
Lend an ear to our screaming shout,
please take us home and let us out.

OUTPOST DAGMAR

By PFC Frank Youngerman, USMC
Item Company, 3rd Battalion, 7th Marines
February, 1953

There was no moon on a very, very dark night,
When the Chinese came out to fight.
They came across the paddy and up the slope,
To live for tomorrow is your only hope.

Up the hill they move, at first kind of slow,
Through artillery and mortars, both friend and foe.
Now, just for a minute you think about home,
Across a wide ocean of high wave and foam.

You dream of your girl Mary, Jane or Joyce,
She's the one you love, the girl of your choice.
Sort of makes you ask if this is the end; or
If, God willing, you'll see her again.

There's no time to daydream, they're moving up fast,
One of em drops with each muzzle blast.
The time goes slow, it seems to drag by,
You'd think it was day, for the flares in the sky.

The enemy is breaking, they start to run.
We've given them hell, and prayed for the sun.
Five hours ago they began the attack,
But if they took Dagmar, we'd have taken it back.

It's just a small ridge, forward of our M.L.R.
Well protected with rifle, machine gun, and B.A.R.
Behind each weapon there's a grimy Marine.
He's the damnedest fighter I've ever seen!

Their clothes are muddy, tattered and torn,
The men are weary, tired and worn.
There's one to your left, and one to your riqht.
You can see them real plain by the dawn's early light.

They're a great bunch of guys; they know how to fight.
As we'll show them again, if they come back tonight!

PFC ROBERT T. OLWINE JR., US ARMY
Nokomis, FL

"My Korean War Experiences"

I SERVED IN THE HEAVY MORTAR COMPANY 19TH INFANTRY Regiment, 24th Infantry Division. On July 30, 1951, I left the United States from Pier 91 at Seattle, Washington aboard the USS General John Pope. Aboard were approximately 4800 men, including 25 army officers, 300 air force officers, and the ship's crew of 275.

From the time we left the west coast of the United States and arrived in Yokohama, Japan, we had gained 17 hours in time. There was a difference of 14 hours between Japan and Ohio, where I was from.

The boat docked in Yokohama on August 10, 1951. We boarded a Japanese train and traveled about 35 miles to Camp Drake — a processing camp for all troops arriving and leaving the Far East.

At Camp Drake we had some chow and started through processing, which lasted until 4:00 a.m. the next morning. At that time, a group of us marched to the theater where we viewed a movie to better prepare us for Korean combat duty. We were taken to the rifle range where we zeroed in our individual weapons before returning to camp.

At mail call we received all the mail that had caught up with us. After that, we exchanged our dress uniforms for more combat equipment and had a clothing check. After chow our names were read designating our division and the time of departure.

I was assigned to the 2D Infantry Division and left by train after spending only 36 hours at Camp Drake. I was taken back to Yokohama Harbor where I boarded the USS Marine Lynx and was on my way to Korea.

I don't recall how many troops were aboard this ship, but it was overcrowded. There were 200 men, including men living on the deck of the boat. We went the whole trip without a hammock in which to sleep.

I arrived at Inchon, Korea, on August 14, 1951, but since the tide was out, we went ashore in LST's. I boarded a train which would take me to the camp where I would receive the final phase of combat training. I'll never forget that train ride because there were no seats in the cars, only bunks of wooden planks. The train ride lasted about seven hours. During this ride, I had my first taste of C-rations. Although it may sound unbelievable, they tasted better than some of the other Army chow I had eaten.

At 11:00 p.m. the train arrived at Ascom City, Korea, the replacement depot for the 24th Infantry Division. A group of us disembarked and the train continued on its way. We marched about 45 minutes with full field pack to our sleeping quarters which turned out to be a bombed-out factory. This is where I saw the first disasters of war, as far as destruction to buildings and property was concerned

We slept on a concrete floor the remainder of the night. The next morning we were processed and permitted to make out last wills and testaments, additional allotments, and other legal documents, if we desired. While I was there, I spoke with a South Korean boy who was walking guard between our building and a barbed wire fence. It was designed to keep the South Korean people away from us and vice versa. I hadn't had any beer since I left the states, so I bought a can of Schlitz beer for $1.00 and a copper ring with carving on its face for $1.50.

After two and a half days in these quarters, we were transferred to squad tents in which about 24 men were sheltered. After a couple of days, I was ordered to pack all my equipment and march to a training area about two miles away for a five day training course to prepare for combat duty.

We took long hikes, had classes on military intelligence studies, enemy weapons, had bayonet training, and fired our individual weapons for accuracy. Our last night at the training area was spent on an all night problem of patrolling. The next morning we returned to the main camp and were given a leisure day.

Then we were loaded on two and a half ton trucks and shipped to our respective units. I went to the 19th Infantry Regiment. It was a five hour ride from Ascom City to the Command Post located in the reserve area, near Chunchon, Korea. The 24th Division was in reserve at that time. I arrived at regiment at 3:00 p.m. on August 24, 1951 and was assigned to Heavy Mortar Company.

I boarded the truck which transported me to the company's reserve area. I was interviewed by First Lieutenant Ryan, the Company Commander, and placed in the Second Platoon. I was interviewed by Platoon SFC Alligood, and put in the Fourth Squad. Of the 180 men I was with in basic training at Camp Breckenridge, KY, there were only six of us still together.

My squad leader, Corporal Maguire from New York City, was on R & R in Japan at this time, and Corporal Psygoda, the Assistant Squad Leader and first gunner, hailing from St. Louis, MO was attending NCO school. PFC Fisher from Wapakoneta, OH, the second gunner in the squad, greeted me and helped me put up my pup-tent.

The 57th Military Police Company was located directly across the road from us and movies were shown there every night. We bathed and washed our clothes in a nearby river. Our daily schedule of physical training and field classes was mostly mortar training where I learned the fundamentals of the mortar and its sight. We had training on laying in the guns at night using only the sight lights and the small light in the aiming circle.

I learned that it was more convenient to transport this weapon by vehicle instead of by hand since it had a total weight of 340 pounds (Base plate—175 pounds; Tube—105 pounds; Sight—7 pounds; Standard—53 pounds).

On August 27, 28, and 29, 1951 we moved up to the Kansas Line and dug mortar positions and bunkers. This line of defense was supposed to be as impregnable as the Siegfried Line was in Germany during World War II. It didn't have concrete pill boxes, but positions were well protected with sand bags. Tanks would have been unable to move through such mountainous terrain.

In the afternoon of September 8th, we were given marching orders. Everyone thought we were going on line. We pulled out of the area around 5:00 p.m. in a convoy and drove for about six hours. We had to drive with blackout lights. We were only 2000 yards from the front line. Then we took the wrong road and lost two hours of time because several of the vehicles ran off the road into a deep ditch. We finally reached our assembly area at 4:00 a.m. and slept until 7:00 a.m.

The morning of September 9th, the First and Third platoon moved up to fire on the enemy. Their purpose was to give support to the Third battalion and get part of the 25th Division out of an area surrounded by the Chinese. Our riflemen met no resistance and our platoons didn't fire a round, although a medic from our company was hit by incoming enemy artillery. When our riflemen reached the hill where our men were being held captive, only one remained alive. Nearly 170 enlisted men and officers were found dead.

Both platoons returned to the company area September 10th and the Company left the area the next day. We drove to Pochon, Korea, where we destroyed a farmer's soy bean field while putting up our pup-tents. One of the Company's house boys, called Joe, said his parents lived just ¼ mile from there.

The following morning we traveled to Chip-o-ree which was another reserve area like Pochon. We had movies in the evening but lots of daily mortar training. We had a firing problem in a rice paddy and another on a firing range.

On October 5th, we left Chip-o-ree and drove to an area about five miles from the front lines. We slept that evening, and the following morning at 5:45 a.m., we went on the front line in the Kumwai area. We were set up behind a large hill that offered us ample protection from enemy artillery.

Four men from the squad pulled all night guard on each squad's mortar. The rest of us guarded the telephone and the 50 caliber machine gun. The gun was placed across the road on the forward slope of another hill. At our position we had only a few incoming artillery rounds. They didn't hit very close to us so we had no casualties.

On the night of October 10, 1951, Shepherd, from the Third Squad heard a machine gun chattering while he was on guard. He set his own deflection, elevation, charges, and fired a mortar round. It silenced the machine gun but had any of the 'brass' found out about this incident it would have meant a Courts-Marshal.

On the morning of October 13th, our riflemen jumped off on the attack, and we moved to a new position where we had very little protection. The following morning at 4:00 a.m., we had some incoming artillery fire. Many of the rounds landed close to us, but again no one was hit.

On October 14th many prisoners were brought back as well as many of our wounded men. Some of the prisoners were seven feet tall. It occurred to me that they were probably Mongolians.

In the afternoon, two tanks which had been knocked out by the enemy were repaired and pulled out. Also on that day, a ¾ ton truck from the third platoon hit a mine and blew up, killing the Squad Leader and injuring all the men on the truck. Five of the men received stateside wounds that required treatment back in the states.

We pulled out of that position on the evening of the 14th and went to a new position. We were in an area where mines were buried but we didn't hit any. We didn't get any sleep at all that night and left the next morning for a new position. We set up the 81 mortars beside us. In a nearby Korean house we found a dead Korean woman and a Chink. Another dead Chinese was lying in a small nearby stream.

When we left that position on the 17th, the truck of the Second Squad of the Second Platoon was blown up. Although the driver received just a couple of minor scratches, a rifle company replacement, who had never seen combat, was killed. After losing the truck, the Third and Fourth squads were moved into position by jeep instead of the squad's trucks.

Our new position was very poor because we were set up on the side a hill instead of behind it. The Chinks had the area well zeroed-in and enemy artillery came in regularly on the road leading to this position. Enemy shells came in on the 18th and proved that we were in a poor position. Shorty Napolitano was killed. Dinsmore, Nesvik, and Shephard were injured.

During most of the day an enemy sniper irregularly shot at us. He shot at our panel (flag) man as our attacking rifle-men advanced up the finger of a hill east of Hill 435.

Men from our mortar pool came out to salvage what they could from the demolished Second Squad truck. While they were working, a bulldozer ran over another mine about five feet away. Two men were injured, one seriously. Our platoon then moved behind the hill that was beside us. This prevented the enemy from using direct fire weapons on us.

Bob standing and sitting at his bunker in 1951

A Chinese soldier surrendered to us on the morning of the 19th. We gave him breakfast but he ate very little. We believe he was the sniper who had been shooting at us. We moved again that evening and the following morning set up in another position located in a draw behind Hill 770. We had to carry ammunition about 100 yards up hill to the guns in this position. When our bunkers were completed, we again changed position.

Our new position was located at the boundary of the regimental sector of control. There was no road, so we hand carried the mortars ¾ of a mile up the hill. We didn't fire that night but guard duty was extensive because we had no other units near us for partial protection. On the morning of the 24th, we hand carried the mortars half way back down the hill and set up again. Then we carried our ammunition ¾ of a mile to this position.

We pulled guard with a captured Chink automatic rifle because it was the only automatic weapon that would fire. Our 50 caliber machine gun would only fire about three rounds before jamming, and our 30 caliber machine gun was so rusty it wouldn't fire. While in this position, Joe Miller shot a pheasant with a captured Chinese rifle. It was awfully good eating so we tried to shoot more of them without success.

On the 4th of November, I was recruited as the radio operator for our forward observer. It was raining and a long climb up Hill 770. When we stopped to rest, about half way up, I handed my helmet to the forward observer, Sgt. Story. He dropped it and back down the hill it rolled so I had to go after it, of course. We reached our objective that afternoon. We were with L Company, and Captain Bell was the company commander.

On the fifth we went to an outpost of the company and fired on a mortar and machine gun. We fired for three hours and didn't see a thing we had destroyed but we sent back a report that we had knocked out an enemy mortar, silenced a machine gun, and killed two dozen Chinese. The next day was too foggy to do any firing.

The seventh, we went back to the outpost to fire again. I didn't carry my carbine because it was very unhandy while carrying a 300 radio on my back. On the way back, we were shot at by an enemy sniper when we stopped to pick up a burp gun. Story got scared and ran off leaving me alone and unarmed but I made it back without any further difficulty. We didn't go to the outpost on the eighth, but I witnessed an enemy ammunition dump being blown up by our tanks.

While on the hill we ate C-rations and pulled guard duty for six hours each night. Every time a twig snapped someone would throw a grenade. We were expecting another counter attack on our position because the Chink's first counter attack on the ninth had been repulsed by us.

We were relieved on the ninth and sent to our platoon. Meanwhile, it had moved forward to a position farther into the draw behind Hill 770 where we were before. We went back to our company rear after chow that evening for some rest but returned to the platoon on the 15th. We stayed in that position until the 22nd which was Thanksgiving. At that position there was a cable that took chow and equipment up to the top of the Hill 770, and it also brought down the wounded.

Thanksgiving we moved to our last position in the Kumsong Valley. We had a big dinner of turkey with all the trimmings. We also had turkey dinners on Christmas Day and New Years Day.

On December 25th we had a time on target fire mission of 65 rounds per gun. We had our own artillery coming in on us on December 30th because one of our 155mm Howitzers had set their elevation wrong. January 1, 1952, we had another target fire mission of 60 rounds per gun.

We left that position at 6:00 a.m. on January 19, 1952. We rode on two and a half ton trucks to the rear area of 2nd Battalion Headquarters. We joined a convoy and headed for Tent City located 12 miles south of Mortar Junction. (Our first position on line had been only 300 yards from Mortar Junction.) We stayed in squad tents at Tent City and attended a dedication service on the 20th, for all the men from the 24th Division who had lost their lives in Korea.

We left Tent City on the 22nd by truck to Chunchon, where we boarded a train for Inchon. We arrived at Inchon the morning of the 23rd where we transferred to LST's and then boarded the USS Noble. At last we were out of Korea and on our way to Yokohama, Japan.

On January 29th we docked in Japan, boarded a train and arrived at Camp Haugen on January 30, 1952. We marched to building five which became our new home. Misumisawa was the name of the town outside of check point one but it was more commonly known as the Gulch. Takadata was just outside of check point two and Hachinohe was about seven mile east of camp. Misawa was about 15 mile northwest of camp and Jim Homan was stationed there at the Air Force base.

On February 26th we had a 30 inch snow storm that lasted three nights and two days. We had training cycles that lasted for 17 weeks, and at the close we always had a maneuver. We had air training during the last part of May and made a 55 minute flight on June 7, 1952.

Our company went on maneuvers July 14, 1952, at Sekine. The second platoon did the best firing and number two gun did the most of it. I left Sekine July 18th and the rest of the company came back the following day.

I left Camp Haugen on July 21st for Sendai to practice for the Division Show, "Readin', Writin', and Rhythm." The show opened August 13th at Camp Schimmelpfennig. We played at Camp Haugen on the 15th and all three camps at Fujo on the 17th, 18th, and 19th.

We left Camp Fujo the morning of the 20th and stayed in Tokyo for nine hours. On the 21st, the show was put on at Camp Younghans and again at Schimmelpfennig on the 22nd...Our last performance was at Camp Sandai on the 26th, then I returned to Camp Haugen on August 27, 1952. I was placed on special duty with special services and was made the light technician of the Regimental show "Draft Tease." This show was put on at Camp Haugen, September 11th and 12th of 1952.

On the 13th of September, I went back to my regular duties in the company and that day we had wet-net training for the amphibious landing maneuver we were to make in the near future. I boarded the USS Bellatrix on the 21st as part of the advance party for this maneuver. We had a practice run on the 28th and made the landing the next day at Chigasochi Beach.

We drove to Camp McNair which was to be our new home until we completed the Regimental Combat Team Test. We had a practice run for the test from October 13th to the 15th and took the test from the 20th to the 24th.

On October 27th, we left McNair by two and a half ton trucks and drove to the beach. Here we loaded on LST's and went out to where the USS Mountrail was anchored. We boarded and were soon on way back to Camp Haugen. We landed there on the 29th at White Beach, north of Tagatay Ridge.

I was pleasantly surprised on November 14th at 9:05 a.m., to find out I was going home via rotation. I cleared the post on November 14th and 15th, 1952. I left Camp Haugen November 16th at 8:40 a.m. by train, and arrived at Camp Drake on November 17, 1952 at 9:30 a.m. Most of my processing was done on the 18th and 19th and I moved to the shipping building on the 20th. I left on November 24, 1952 at 8:15 a.m. and traveled by bus to Yokosuka where I boarded the escort aircraft carrier, USS Windham Bay.

On December 9, 1952, at 11:00 a.m. the ship docked in Almeda, CA. I was home again after spending 16 months and ten days overseas, including 49 days aboard ship.

I received the following letter in 2001 from the President of Republic of Korea, Kim Dae-jung.

June 25, 2000

Dear Veteran

On the occasion of the 50th anniversary of the outbreak of the Korean War, I would like to offer you my deepest gratitude for your noble contribution to the efforts to safeguard the Republic of Korea and uphold liberal democracy around the world. At the same time, I remember with endless respect and affection those who sacrificed their lives for that cause.

We Koreans hold dear in our hearts the conviction, courage and spirit of sacrifice shown to us by such selfless friends as you, who enabled us to remain a free democratic nation.

The ideals of democracy, for which you were willing to sacrifice your all 50 years ago, have become universal values in this new century and millennium.

Half a century after the Korean War, we honor you and reaffirm our friendship, which helped to forge the blood alliance between our two countries. And we resolve once again to work with all friendly nations for the good of humankind and peace in the world.

I thank you once again for your noble sacrifice, and pray for your health and happiness.

Sincerely yours,

Kim Dae-jung
President of the Republic of Korea

TSGT LEO G. RUFFING, USAF
Portsmouth, VA
National Chaplain,
Korean War Veterans Association

I QUIT HIGH SCHOOL AT AGE 17 AND ENLISTED IN THE ARMY. After completing basic training at Camp Breckenridge, KY, I was assigned to the 11th Airborne Division on occupation duty in Japan. When the 17th Regiment of the 7th Division came from Korea to replace the 11th Airborne Division, I transferred to the Heavy Mortar Company, 17th Infantry Regiment during the Army Occupation of Japan. We were stationed at Camp Schimmelphennig in Sendai, Japan.

I returned to the States in March of 1950 just before the Korean War started in June. I was in the active reserve and applied for active duty. They assigned me to train basic soldiers at Indiantown Gap, PA. I volunteered to go to Korea and requested that I be assigned to the 17th Regiment.

It took some time for the transfer to go through and in 1951 I returned to Korea but was assigned to "M" company 32nd Infantry Regiment which was in reserve in northeastern Korea. My participation in the war was rather uneventful.

Leo G. Ruffing and his youngest son, Michael, taken during their visit to Seoul in 2008 having lunch with the President of the Republic of Korea, The Honorable Lee Myung-bak.

In late October 1951, my first Sergeant advised me that I could leave the unit and return to the United States for discharge from the army. I left the unit and caught a ride on a weapons carrier to the 7th Division Headquarters located at Chunchon, Korea. A brief delay while they completed the paperwork and I was instructed to take a train to Inchon and board a troop ship for the ride home.

Early in the morning as I boarded an almost empty car on the train to Inchon, I noticed a Corporal entering the car. There were many empty seats in the car but the Corporal looked around and walked over to the seat next to me. Then he extended his hand and said, "Hello, my name is Pete."

I learned his name was Peter A. Petrulo, and he was also being discharged. We were both from Pennsylvania. I was from Pittsburgh and Pete was from Butler, a few miles away. We talked all the way to Inchon. Pete was a baker and served with mobile units in North Korea. He personally had not engaged in any battles except the one against hunger.

We arrived in Inchon, and were told that during the night a British freighter had drifted and knocked a hole in the side of the ship that was supposed to take us home. We got back on the train and made the long, slow trip to Pusan, in South Korea. Trains were very slow in those days. Arriving in Pusan, we were placed in "tent city" and told we would have to wait until a new ship was assigned to take us home.

Then the typical Army "snafu" kicked in. After a few days waiting, we were ordered onto a ferry (very small and very crowded) that carried us to Sasabo, Japan where we waited in another, tent city for about a week. We were told that a ship had been located and we would begin our voyage home the next day. Lo and behold we boarded the ship that was originally scheduled to take us home from Inchon. That's right, the ship with the hole in the side.

At Yokuska they welded plates on the side of the ship making a lot of noise. They worked night and day and it was impossible to sleep. It was a good thing they had taken our weapons away when we left Korea, or there might have been a shoot out between us and the Navy!

The trip home was a totally different story. The ship was over loaded and short of rations. We ate our first meal at six AM and another meal at six PM with nothing to do in between except complain about how slow the boat traveled and bemoan showering in salt water. Bunks were canvas and stacked one above the other.

All the way across the wide Pacific, Pete and I hung out together and shared our dissatisfaction with the way we were being treated on the way home. As we arrived at Port Mason in San Francisco, I remembered the GI's returning from WWII with all the excitement, bands and parades.

When we arrived at Port Mason after dark, there were no parades, bands, or excitement. There was one lady in a red dress who sang as we disembarked. Except for her, no one noticed that we returned. We loaded onto buses and were transported to Camp Stoneman where we were processed, paid, and ordered to our final duty station.

Pete and I were given train tickets and ordered to report to Indiantown Gap near Harrisburg, PA. I guess that is as close to home as they could get us. We remained there during our last few days in the Army, and then we were discharged on 13 December, 1951. Pete and

I said goodbye with the unspoken understanding that we would never see each other again.

I went home to Pittsburgh and within a few days found a job. It was about two or three weeks later that my brother, Jim, my youngest sister, Alice, and I were walking down one of the streets in a part of Pittsburgh called Homewood. Lo and behold, we ran into Pete. So I introduced Pete to Jim and Alice. Somewhere in the conversation Pete was invited home for supper. We all went home for supper, and Pete never left.

On September 6, 1952, Pete and Alice married and stayed married until Pete's death on January 1, 2002. Between 1954 and 1965, Pete and Alice had six children, three boys and three girls, who are all great adults at this writing. Pete was a treasured member of our family for 50 years. We were friends as well as in-laws and never exchanged a cross word.

Pete told me the reason he did not stay in the military was because he did not want to travel far from Pittsburgh. Pete took advantage of the GI bill and worked his way through college, earning a BA from Duquesne University and attending graduate school at Western Michigan University.

He made a career in the food service industry and worked in a variety of places. During Pete's career he managed companies in Paw Paw, MI; Kennewick, WA; La Porte, IN; Wheaton, IL; Duluth, MN; and St. Paul, MN. Not bad for a guy who wanted to stay close to Pittsburgh and did not want to travel. It was right and fitting that Pete was buried with full military honors at Fort Snelling, MN.

I remained a civilian for a short period of time, but in February, 1952, I joined the United States Air Force and made a career in the military. When I enlisted, I was sent to Wright Patterson Air Force Base in Dayton, Ohio where I became a military policeman. At this point I started having some real problems with alcohol. I faced a military court three times during 1952, and spent some time in the base stockade.

In early 1954, I was arrested for a DUI and thought I would be put out of the military. They gave me a break and let me volunteer for reassignment to Korea. I spent 13 months in Korea and managed to stay out of trouble despite the fact that I continued having trouble controlling my use of alcohol.

For the next three years I had several close calls with the law because of drinking. At one point in 1957, I was involved in a pretty bad automobile accident and spent some months in a military hospital. While hospitalized, I found a solution to my alcohol problem.

I returned to duty and I have not had a drink of alcohol since that time. I completed high school, college, and started work on a Masters degree. I retired honorably from the USAF in 1970.

After earning my MS degree, I entered Seminary. I completed a Doctorate in Christian Ministry, and following my ordination, I served a national religious organization for 11 years. My ministry was devoted to working with people who had problems with alcohol and/or drugs.

Some years ago the Department of Military Affairs, Commonwealth of Virginia, reactivated the Virginia Defense Force. This is an auxiliary unit of the Virginia National Guard. I was commissioned and assigned as Chaplain for the 2nd Battalion of the Lafayette Brigade.

When John Edwards was elected State Commander of Virginia's Korean War Veterans Association (KVWA), he appointed me a State Chaplain.

Upon the retirement of Chaplain Len Stegman (Col. US Army Ret.) from the position of National Chaplain, I was appointed National Chaplain by the President of KWVA Lou Dechert (Lt. Col. US Army Ret.). It has been my pleasure to be able to serve the men with whom I served so many years ago. Respectfully, Leo G. Ruffing, National Chaplain (KWVA)

1STSGT LEON B. WARD, US ARMY
Palm Coast, FL

I WAS BORN IN BROOKLYN, NY ON FEBRUARY 24, 1932. I ATTENDED public elementary schools and high schools while living in Brooklyn. In 1948 I joined the National Guard and learned how to cook. They sent me to Fort Devens, MA, for further training.

While there, I experienced my first time in jail. One of my fellow service members decided that he was going to get his watch back from one of the ladies he was seeing, and a fight broke out. Someone called the police and a paddy wagon pulled up to the scene. Yes, you guessed it; the first person they placed in the police wagon was me.

They had us take off our belts and shoe laces so we wouldn't harm ourselves. The next morning we were released and sent back to our unit. Nothing else ever happened regarding that incident.

In 1950, I decided to enlist in the Army. My great grand parents had to sign for me. I was stationed at Ft Dix, NJ, where I went though 12 to 13 weeks of training. It was fun at times, but not so much at other times. There were trainees from all over the Eastern part of the United States.

At that time we were segregated from the White soldiers. They managed to have White officers commanding us, and other colored service personnel training us. Our barracks was located near the firing ranges. A fence was placed between the lady soldiers in the Womens Army Corps, and our area.

Shortly after the Armed Forces were integrated, I was assigned to a Quartermaster Unit in support of the troops serving in Korea. Our

Leon Ward (Left) and Owen Johnson

unit was located in Japan. We had sixteen personnel assigned to the unit with only two colored soldiers. I received the Korean Medal for being in the theater of operation.

One day, the MP assigned to the unit was drinking a little too much (it wasn't water) and decided it was time to let me know that I was black. At the time I felt sorry for him, because his parents had failed to teach him that using those kinds of words during those times was very offensive.

I attended integrated schools and never had someone call me that. I told this person to just leave me alone. He refused and proceeded to place his hands on me. When he did, I decked him and he fell to the concrete floor.

I was very nervous, thinking I had killed him. Thankfully, I hadn't. Then I went to the commanding officer and reported the incident. The soldier was reprimanded and released.

He had the gall to later exclaim, "Well aren't you Black?" What could I say? During those days, of course, the term was not used. It was either Colored or Negro, but never Black.

Another incident took place in the kitchen. I was cutting up chicken with a meat clever and when I looked up I almost severed my left thumb. They rushed me to the hospital. I guess I blacked out from loss of blood. When I woke up several pretty nurses were standing around me, and hovering over me. I thought I was in Heaven, but was ever so glad to be alive.

Not all of my time in the military was good, but most of it was. The best part is that the government is still sending my retirement checks. My advice: when you are making up your mind to do something always think about your future, and where you want to be.

During my military career, I was stationed in Kentucky, Japan, Germany, Vietnam, and Niagara Falls, NY.

1STSGT FREDERICK W. WORRILL, US ARMY
Crawfordville, FL

I ARRIVED ON OKINAWA IN FEBRUARY, 1949 AND ASSIGNED TO B Company 44th Infantry Philippine Scouts which, in June 1949, was renamed the 29th Infantry Regiment. I was trained and qualified as a machine-gunner. Later, when the First Sergeant found out that I could read, write, and type, I was assigned to the Supply Room as the Supply Clerk. I did not mind that at all.

The Korean War started on 25 June 1950 with the North invading the South. Around 15 July, the 29th Infantry Regiment was alerted for deployment to Korea. We were supposed to go to Japan for six weeks of intensive combat training. That did not happen. On 20 July, the 29th departed Okinawa on the Tagasaki Maru, a former Japanese hospital ship.

We stopped in Fukuoka, Japan overnight to refuel and then sailed straight to Pusan, Korea. From Pusan, the 29th moved to Chinju and set up in a large school building. The following day, 27 July, "B" Company moved out to Anui to relieve "A" Company, 19th Infantry Regiment, 24th Division. The relieving action coincided with an attack by the North Korean Army. "B" Company sustained approximately 62% casualties. In September, 1950, the 1st Battalion, 29th Infantry was re-designated as 3rd Battalion, 35th Infantry, 25th Division. "B" Company became "K" Company.

After the Chinese entered the Korean War and overran our positions in November, 1950, we received orders to withdraw. As we worked our way along a ridge-line parallel to a road, we saw and heard a GI calling to us to come down. We thought it was one of our Lieutenants. We had already stripped our machine gun and had thrown the parts into a creek that was partially iced over.

As we got to the road, we were suddenly surrounded and taken prisoners by Chinese soldiers, several of whom spoke English. We were herded into the yard of a Korean house where they gave us hot water seasoned with what I thought was black pepper. It seemed to warm me up.

The soldiers marched us toward the North that night. Just before daylight we stopped at a small Korean Village and were hustled into a house. Exhausted, we all went to sleep. I was awakened by a thunderous roar as American jets strafed the village. Almost instantly, I felt a tremendous jolt and extreme pain as the armor piercing core of a 50 caliber bullet slammed into my left side.

Knowing what I had been hit with made me think I had been cut in half. My Gunner, Alvi Norris, who was lying beside me, tried to help. I told him not to move me because I thought I would tear apart. The Chinese guards made everyone who could walk leave the building to take cover outside and I did not see them again. I was the only one wounded in the strafing attack who survived. There were three or four others wounded, who did not survive.

The Chinese could have just left me there to freeze to death in a day or two, or for North Korean soldiers to come along and kill me. Instead, they gave me a GI sleeping bag and put me in with their wounded comrades and took me all the way to Manchuria, to a railroad tunnel on the other side of the Yalu River.

I lost track of time, but after I was there for a period of time, a Chinese officer came along and discovered me mixed in with a bunch of Chinese. After a lot of discussion and arm waving, they laid me on a ledge on a railroad tank car with two Chinese soldiers to hold me on.

This was shortly before Christmas, 1950. They sent me back into Korea over a railroad at the Hydro-Electric Dam. This place I believe was named Supong, a fairly good-sized city with brick buildings and a two-story hospital. Getting to the hospital was an ordeal in itself.

Some nights we traveled on mule carts, some nights by truck, some nights by train, and some nights by improvised litter with four labor troops carrying one litter. All the while, we had to dodge the US Air Force. During our stops at daylight, we hid in farm buildings, in the woods, and in railroad tunnels. After arriving at the North Korean hospital, the Chinese soldiers left me, and I was put into a room with three other wounded US soldiers.

One day a truck came through town carrying Air Force personnel to the POW Camp #1 at Chang Ni. They stopped the truck and we were hustled out, put on it, and sent on our way to Camp #1 with them. We estimated it was about a 35-mile trip. It was a good move for us because it got us into the POW system.

I was carried as an MIA (Missing in Action) for a year. My family was notified that I had been listed as a POW by the Chinese just before Christmas, 1951.

After my release, I read the account of the US Air Force bombing of the Hydro-Electric Dam on the Yalu River. I am pretty certain it was the dam in the town where we were held prisoner at the hospital.

From POW Camp #1, we counted 64 aircraft headed toward the Yalu River about the time the article mentioned. I think it would have been a very dangerous situation for us had we been there at the time of the attack. More likely, we would at the very least have been beaten, if not killed.

I believe it was in September, 1951, that we were sent to Camp #1 where, after a few days we were separated by rank. The officers were kept across the road from the enlisted men who were separated by rank, with the NCOs on one end of the camp and lower ranks on the other end. There were American and British soldiers together with a few Turks mixed in.

Later, I think in 1952, the Chinese established a Sergeant's Camp four on the Yalu River at Wiwon where we were further separated. White Americans were put in one area with a stone wall between us; and the British and Turks were put together; Black Americans were in another area.

This is where I was on July 27, 1953, when the Armistice was signed. I was repatriated in September 1953, the same day as, and about 15 minutes before, Major General Dean was released. I had the honor of standing about 10 feet from the General when he was presented the Korean Medal of Honor by Sigmon Rhee.

I was evacuated back to the States and assigned to the Detachment of Patients at the US Army Hospital at Fort Gordon,

Worrill — Last tour of duty

Georgia. While I was a patient at Fort Gordon, they discovered that the 50 caliber bullet had not been found when the Chinese performed surgery in 1950.

I underwent surgery at Fort Gordon and the bullet was removed from my leg joint where it had stopped after passing through my pelvis. The bullet still had bits of mud on it. Surely, God Himself was watching over me.

After being discharged from the Army, I took a short break in service to marry the love of my life. We are still sweethearts 54 years later.

I re-enlisted in the Army, keeping my rank of Master Sergeant and completed over 20 years, retiring as a First Sergeant in the Artillery in 1969. After retiring, my family and I returned to our hometown of Savannah, Georgia.

I had several civilian jobs, but I found my niche in 1973 by getting back into uniform. My first job was a Jailer/Booking Guard with the Savannah Police Department, and then I went with the Savannah International Airport Police Department where I retired as a Captain in 1992. After my second retirement, I had several part-time jobs like delivering and picking up film, delivering auto parts, and inspecting and photographing real estate.

Our oldest son's work had taken him to Tallahassee, Florida, where he married a Florida girl in 1989 and started a family. In 1998, we moved from Savannah, Georgia, to Crawfordville, Florida, to be near our two grandsons, and help out when needed.

In 2005, I had open heart triple by-pass surgery. So now I am really retired with a 100% disabled rating by the VA due to the after effects from my POW days.

MAJ GENERAL FREDERICK C. 'Boots' BLESSE, USAF
Melbourne, FL

GENERAL BLESSE WAS BORN INTO A MILITARY FAMILY IN COLON, Panama Canal Zone on August 22, 1921. His father was a Brigadier General in the Medical Corps of the US Army, who retired in 1953. Young Frederick Blesse graduated from American High School, Manila, Philippines Islands in 1939.

He graduated from the US Military Academy, West Point, NY in June 1945 commissioned as a Second Lieutenant, with a Degree in Science and Engineering. After graduation, Blesse chose to enter the Army Air Corps. He earned his rating as a pilot and spent 28 years actively flying Fighter Aircraft.

During the Korean War, he flew two volunteer combat tours. Between November 1950 and May 1951 he completed 67 missions in the P-51 Mustang aircraft with the 67th Squadron, 18th Fighter-Bomber Wing. In February 1951, he was transferred to the 7th Squadron of the 49th Wing where he flew an additional 35 missions in the F-80.

With his first tour complete, he returned to Nellis AFB. He volunteered to return for a second combat tour in Korea and in March 1952 he was reassigned to the 334th Squadron, 4th Fighter Group where he flew 123 missions in the F-86.

During his second combat tour, he was officially credited with destroying nine MiG-15's and one LA-9 aircraft; probably destroying one additional MiG-15 and damaging three other MiG-15's. He was the leading jet ace in the US Air Force when he returned to the United States in October 1952.

General Blesse's Korean War experiences include his crashing a P-51 Mustang under extremely low visibility conditions, into a truck that illegally crossed a short North Korean runway. Below, in his own words, he describes how he ejected from an F-86 after running out of fuel while shooting down one MiG-15 and damaging another:

"On 3 October 1952, during my 123rd mission in Korea, my wingman and I cruised around looking for MiG's, but didn't see any. As we turned around and started South, we saw a flight of four MiG's at our six o'clock, out two miles. They were above us and closing. We didn't have the fuel to fight so I wanted to get out over the water and continue home. The MiG's attacked and I thought, "Christ, we don't have fuel for this." Nevertheless we fought.

During the fight, with low fuel, I told my wingman to get over the water and head for home. As frequently happened during those damn fights, I looked around after pulling out of a roll and found I was all by myself. I knew two MiG's were there someplace, but I couldn't find a thing.

I stayed in a right turn for about ten seconds, worried somebody was going to take shot at me. Then I looked at my fuel gauge— 1,100 pounds. If the winds were right I'd make it home with fumes, but if they were wrong, I'd have to try to land on the beach on Peng Do, a small island off the coast. I started climbing to save fuel.

Around 11,000 feet, I caught a glimpse of something out of the corner of my left eye—a MiG, coming down on me from 10 o'clock. "This is it." I thought, "School's out. I don't have any fuel to fight with this guy." For some reason he never saw me, diving right in front of my F-86. As he went by, I looked at my fuel gauge, then looked at the MiG and said, "Hell, it's a tossup anyway. Why not?"

Pulling about a 4-G turn, I climbed to the right and rolled off the top underneath him. It took about 25 seconds—I closed to 600 feet, fired and watched the MiG explode and begin burning, then saw the pilot eject. I rolled right back out on course, figuring it cost me about 200 pounds of fuel to shoot him down. There went Number Ten but I was down to 900 pounds. I hoped the camera film worked, otherwise, no confirmation of my tenth kill.

The alert flight at Kimpo Air Base, led by Captain Robbie Risner, had been scrambled hearing we were in a fight and short on fuel. It was a good thing Robbie and his flight saw the burning MiG-15 or I wouldn't have been given credit for my tenth kill. I didn't think I would make it back and my film would be lost with the aircraft.

I was still over North Korea when I called Dumbo, the rescue flying boat that was always on station during our missions. I said, "Orbit Peng Yeng Do Island, because If I make it home that's where I'm going to try to land." I thought that If I could get 100 miles out at 39,000 feet with 300 pounds of fuel, shutting down the engine would allow me to glide back to base at 180 knots, putting me in the traffic pattern at 6,000 feet where I could restart and land.

But I only made it to about 32,000 feet when I hit 300 pounds. Without tailwinds there wasn't a chance. Calling the base, I asked what the winds were and they replied, "Out of the Southwest." Nothing could have been worse. I shut the engine off and began a 180 knot glide toward Kimpo Air Base.

As I passed 17,000 feet, I decided I'd become a POW if I stayed on this course. I had no relish for that especially after Pyongyang

Sally had broadcast the night of my eighth kill, "Just wait, Major Blesse, we're going to get you and when we do, we're going to hang you from the Han River Bridge." I changed course and continued my glide toward the west coast and the island. I called Dumbo again, "I'm probably not going to make it, so orbit between the shore and the island." He acknowledged.

I continued to glide down at 180 knots but the wind was against me. For awhile I didn't think I'd even get to the coast. At about 7,000 feet everything looked like it was much too far away so I restarted the engine and used my remaining fuel to climb as high as I could. At 13,000 the engine flamed out so I re-established a 180 knot glide, hoping I'd make the water.

I had to cross a main supply route, and I drew a tremendous amount of enemy fire. Several times the flak bursts were so close one wing or the other would rise up. I had no choice but to be a predictable target—jinking would cost me airspeed and altitude—so I held my heading and hoped I wouldn't get "the golden BB" (a direct hit).

I finally crossed the North Korean coast at about 3.000 feet, but a half mile or so out over the water, I knew I was getting too low. "Time to get out' I thought, so I called Dumbo. "I can't go any farther: I'm getting out. Don't let my aircraft hit you." Instantly he radioed back, "Rog, we've got you in sight."

My hands went to the seat triggers. *BOOM!* Explosive charges blew me and the seat, with parachute attached into the slipstream I went out at about 1,200 feet, a little too low. I didn't want to get tied up with the seat, so I undid my seatbelt before ejecting. The chute opened, pulling me away from the seat instantly. Almost immediately I went into the water. My throat burned in the salt water from a parachute strap abrasion. One of my pockets was open and all my maps and survival gear floated out.

After inflating my one-man raft I swam over and picked up my helmet, which for some reason had not sunk Very carefully I put everything back in the dinghy, then climbed aboard myself. Just about the time I was set for the winter, the Dumbo landed and taxied over to me.

A line was fired out an open window, which I grabbed, and I was pulled over to the airplane and inside. As soon as I was in, the pilot hit the throttles to get out of there. Scrambling to my feet, I ran up the aisle to the cockpit and started banging on the pilot's helmet, 'Wait, a minute! My dinghy is still out there with all my stuff in it." With a quick, disgusted look at me, he said, "Fuck your dinghy. They're shooting at me and we're getting the hell out of here!" My dinghy became a target for Risner's alert flight, which sent it to the bottom of the East China Sea with my F-86.

By the time I got back to Kimpo, word of my bailout had already gone to the Pentagon and a wire had come back which read, "Subject Major Frederick C. Blesse. Subject officer will be returned to the ZI immediately. I was now the leading ace in Korea, and though I suspected it wouldn't last too long, I intended to enjoy every minute of it. The next day I was in Japan for a series of press conferences.

In 1955, 'Boots" Blesse wrote an outstanding fighter tactics book titled, "No Guts, No Glory." This book has been used as a basis of fighter combat operations for the Royal Air Force, Marines, Chinese Nationalist, Korean Air Force, and US Air Force since 1955. In 1973, 3,000 copies were reproduced and sent to tactical units in the field.

During two tours in Vietnam Blesse flew 157 missions in an F-4 with 108 of them over North Vietnam. While in combat in 1967 he was successful in his efforts to convince the Air Force of the necessity for guns on fighter aircraft

'Boots' Blesse finally retired in 1975 after 30 years service including 7,000 hours of fighter time, of which 650 were combat hours. He is the nation's sixth ranking jet ace and is the leading living jet ace today.

After his first retirement as a Major General, Blesse worked 11 years full-time and five years as a consultant for Grumman Corporation.

In 1980, he and his wife, Betty, were rescued by helicopter from the 25th floor of the MGM Grand Hotel in Las Vegas, NV during a huge fire.

Following his second retirement from Grumman, he wrote his autobiography, entitled "Check Six — A Fighter Pilot Looks Back." He currently lives with his wife in Melbourne, FL, where at age 87, he remains fully occupied trying to reduce his golf handicap below 11 which is slightly higher than the five he had 24 years earlier.

Left: 'Boots' Blesse in the air and (right) on the ground.

CORPSMAN BEN "DOC" WOLFE, USN
(Deceased)

*US Navy Corpsman serving with Charlie Company,
1st Bn, Fifth Marine Regiment, during the Korean War;
Submitted by Cpl. John Rick Kennedy, USMC*

"The Straw Hat and the Seersucker Suit"

OVER THE "SQUAWK BOX" CAME THE BASSO-PROFUNDO VOICE of our Gunnery Sgt. "Now hear this, now hear this! Fall in at 0900 in field dress with helmets and web belts." Why wasn't it the usual dress greens?

This is Saturday morning. It is supposed to be a cursory inspection and then liberty. But, "Mine is not to reason why, mine is but to do or die. Semper Fi." Nobody knew anything, but that didn't stop the scuttlebutt, and rumors were rampant.

Men began tumbling out of the barracks and forming ranks on the blacktop. At 0900, our Gunny Sgt appeared. Commands were barked, "AH-TEN-SHUN, DRESS RIGHT DRESS...TWO! COMPANY, AH-RIGHT FACE, FORWARD MARCH, AH-one-twop-ah reep four, Ah-one-twop-ah reep four."

We pulled away from the barracks and with a column left and a right oblique, we were on the road swinging along at an easy cadence of 120 a minute. We passed the PX, went down a slope and past the B-15 Area Theater on our left, and headed toward Basilone Road, named after "Manila" John Basilone from New Jersey and a Medal of Honor Winner on the "Canal".

Still no mention was made of where we were going or why. There were some rumblings in the ranks. Mumbled comments just loud enough for the enlightenment of the surrounding Marines, but not loud enough for the Gunny to hear; "This chicken-shit outfit, why can't they let us in on the skinny."

Others needled each other with breezy wisecracks; O'Neil, Smith get in step." "Stuff it Johnson." "Up yours Mac." "Jones you march like old people f__k." It was friendly banter, something that comes with being a Marine. No summer soldier or sunshine patriot here. Everyone was a volunteer, a raggedy-ass Marine, a veteran of a foreign war. We had claimed the title.

Still in a rhythmic cadence, a left oblique had veered us off Basilone Road and on to a grassy playfield/parade ground. We stepped lively, nay it was a cocky step, with a bit of a swagger that comes from "picking" em up and laying 'em down for many an hour on the grinder. Even the "chancre mechanics" (Corpsmen) marched well.

"Column Right-Hup!" We swung north for a short distance. "Company Halt!" We faced west toward Basilone Road. "AT EASE!" Some distance away, we could hear other "Non-Coms" calling cadence. Three other companies of Marines were marching toward us, coming from different directions.

At least we knew where we were, but why were we here? The other three companies marched up, with various and sundry commands, formed a hollow square. The open area was about 50' x 50' with companies facing each other. Still, no word of what was going on. Being "AT EASE" we were able to turn and speak to one another but the other Marines didn't know any more that we did.

Suddenly, from off to the North, we picked up the faint sound of a drum. As it drew nearer, we were able to discern that it was not just any drum, but the harsh, raspy, grating, abrasive, grinding sound of a snare drum, coming closer and closer, getting louder and louder.

What we saw coming from the direction of the Post Brig gave us pause. Coming toward us was a man in a blue and white striped ill-fitting, cotton seersucker suit. On his head was a fedora style straw hat, pulled down low over his brow. In his right hand, he carried a small, dark brown, hard cardboard 1930's type suitcase, with a canvas strap around the middle to secure it and a "clothesline" rope handle to carry it by.

This sad, wretched, forlorn, pathetic figure had the appearance of a forsaken, hopeless man straight out of the Grapes of Wrath. Directly behind him, about 10 paces back, was a Marine Guard or "Prisoner Chaser" as they were called, carrying an M-1 rifle at high port. Off to the Chasers left, about three paces, was the snare drummer tapping out, *"Rat-tah-tat-tat-rat-tah-tat-tat-rat-tah-tat-tat-tah-tat-tah-tat."* Slowly the three men drew closer to our formation

The snare drum had that grating, grinding, metallic sound of foreboding. In my mind's eye, I was able to conjure up the Place de la Revolution in *A Tale of Two Cities*, and recalled ever so vividly, Madame Defarge knitting into her scarf the story of the hated royalist St. Evremonde family.

At the roll of the snare drum, Madame Defarge would stop knitting, look up, and watch the guillotine fall as the torso of another royalist was shortened by a head. Then she would begin rocking in her rocker and resume knitting until the roll of the snare drum again signaled that the blade was about to fall and behead the next victim.

While they were still 50 yards away, commands were barked to all the companies. "Ah-Ten-Shun! Dress Right-Dress...TWO!" The Non-Com of the company on the North side gave the command, "Dog Company, Open Ranks-Hup!" Two files side-stepped to the right, forming an avenue, an entrance, into the hollow square.

A Major and his aide entered the hollow square and stopped in the middle. The prisoner and his entourage were not far behind. Now right in front of us, the snare drum had a mean, snarly sound. The "Rat-tah-tat-tat-rat-tah-tat-tat-rat-tah-tat-tat-tah-tat-tah-tat" grated on our ears. The "Chaser" gave the command to "Halt!"

The drummer stopped. "Right Face!" the prisoner responded as ordered. A Non-Com barked out "Dog Company. Close Ranks!" The prisoner hung his head in shame. Sweat covered his upper lip. He was surrounded by four companies of Marines. He had a hang dog look about him, but not without reason.

The Major stepped forward so that he was about an arms length from the prisoner. In a loud voice, for all to hear, the Major read the specifications of the Dishonorable Discharge:

> "FOR STRIKING A SUPERIOR OFFICER AND DESERTION IN A TIME OF WAR, YOU ARE HEREBY DISHONORABLY DISCHARGED FROM THE UNITED STATES MARINE CORPS. YOU WILL FORFEIT ALL PAY AND ALLOWANCES. FURTHER, YOU HAVE UNTIL SUNDOWN TO CLEAR THE ENVIRONS OF CAMP JOSEPH H. PENDLETON AND OCEANSIDE, CALIFORNIA. YOU HAVE BEEN PROVIDED WITH MEAL MONEY AND A BUS TICKET TO YOUR HOMETOWN OR PLACE OF ENLISTMENT. IF YOU ARE FOUND IN THE VICINITY OF THE ABOVE MENTIONED AREAS AFTER SUNDOWN YOU WILL BE ARRESTED AND YOU WILL SERVE AN ADDITIONAL YEAR AND A DAY AT HARD LABOR IN THE POST BRIG."

The Major took two steps back and gave the command, "COMPANIES, ABOUT FACE!" All four companies immediately executed an "about face." I thought, "Good Lord, we've turned our backs on this man." How could a man so disgraced ever look in a mirror again? I had heard the phrase, "Drummed Out of the Corps," but I never thought that I would ever witness such a spectacle.

Goose bumps formed on my skin, shivers went up my spine and my pulse quickened. I was actually witnessing a man being drummed out of the Corps. The censuring words of the specifications seemed crushing. What a disgrace! What humiliation! The man was surrounded by shame. I uttered a silent prayer, "Dear God, forbid that I should ever do anything that would cause me to be so dishonored."

It is axiomatic that no Marine will ever fail to go to the aid of another Marine in combat, no matter the cost. It is a trait that is imbued into each Marine during the rigors of Boot Camp. We fight as a team. Each Marine has a job to do. Failure in your task puts your fellow Marine at risk and in harm's way.

We are taught to look out for one another. If we don't, who will? If one cannot count on his fellow Marine, to whom can he turn? It is the bond that differentiates Marines from all other armed forces, except, perhaps the Rangers, Seals and Airborne. This bond, plus training and strictest discipline places Marines in a pre-eminent position among the world's fighting forces.

We are a band of Brothers and this man could not be trusted. The large overhead sign at the entrance to the Receiving Barracks at the Marine Recruit Depot (Boot Camp) in San Diego, says it all:

TO BE A MARINE YOU HAVE TO BELIEVE IN:
YOUR GOD, YOUR CORPS, YOUR FELLOW
MARINE AND YOURSELF. SEMPER FIDELIS.

"Dog Company, Open Ranks — Hup!" Again an avenue opened up on the North side of the square. The snare drummer began his abrasive Rat-tah-tat-tat. The prisoner was marched out of the square. "Dog Company, Close Ranks — HUP!" The Prisoner, the Drummer, and the Chaser headed for Basilone Road toward the Ranch House.

The four companies were marched to their respective barracks and dismissed. We started to speak among ourselves about the tragic melodrama that had just been played out before our very eyes. We entered the barracks, stowed our 782 gear, showered, put on our dress greens and took off on liberty.

A few of us hung back a bit, absolutely overwhelmed by what we had just witnessed. Approaching our Gunny Sgt. to pick up our liberty card, we asked him what would be happening to the prisoner. The Gunny answered that the prisoner would be escorted to the Main Gate.

I said, "Hell, Gunny, that's about six miles away." He nodded in agreement. "What about the poor Chaser and the Drummer?" I asked. He answered, "That's the breaks." Then Gunny added, "The prisoner will be marched to the main gate and brought to a halt right on the Camp Pendleton boundary line."

The Chaser will step up, and with rifle at high port will unceremoniously but with force, give the prisoner a hefty shove in the back, literally throwing him off Camp Pendleton. A jeep will be waiting to give the Chaser and the Drummer a ride back to the brig. I said,

"Gunny, you're not serious are you?" He answered, "Go see for your self."

Pete Simon, from Oakland, owned a Buick convertible and we were planning to go to Burbank and North Hollywood for the weekend. Before we headed north on Highway 101, we drove on Basilone Road to the Main Gate and passed the prisoner and escorts on our way.

By this time they were approaching the Main Gate and had about a mile to go. We didn't have to wait too long before we saw exactly what our Gunny had described, The Chaser called a halt right on the line and then gave a shove that almost sent him sprawling.

My mind harked back to my parochial school education and my high school course in American Literature. I recalled my reaction as a boy, to reading about Philip Nolan in Edward Hale's *The Man Without A Country*. I vividly recalled that Philip Nolan, having committed a military offense, had been sentenced to a navy vessel never to be allowed to disembark.

He attempted to read aloud Sir Walter Scott's poem, "The Lay of the Last Minstrel." Standing on the deck of the ship in an American Port, he was able to see the United States of America, but never set foot on her shores again. With tears streaming down his cheeks, he read the famous line, "This is my own, my native land." The pathos, the deep emotion and anguish of his predicament caused him to choke up and the words stuck in his throat.

SGT MAJOR ROBERT C. WILSON, US ARMY
Cilo, MI

I WAS BORN AUGUST 19, 1930, IN FLINT, MI AND GRADUATED from Flint Northern High School in June, 1948. I started working at the Buick Factory in Flint, MI, in November of that year.

I was drafted into the Army on August 9, 1951 and took 16 weeks of Infantry Basic Training, at Fort Riley, KS, with the 10th Infantry Division. I had a ten day leave and then reported to Camp Stoneman, CA, on December 29, 1951.

I was there about a week and then shipped out to Japan where I attended Army Signal School at Eta Jima. After four weeks of schooling, I graduated as a Teletype Operator. Then I was given orders to go to the 7th Infantry Division in Korea as a Teletype Operator.

When I arrived in Korea, they changed my orders to rifleman with the 25th Infantry Division. The 25th Division was on the front lines in the Punch Bowl, and I was assigned to Company C, 1st Bn, 35th Infantry Regiment, 25th Infantry Division. I served first in the 2nd Squad and then the 3rd Squad, where I remained until I rotated home.

When I went on patrol at night the first and second time, I had to carry the radio. This job was usually given to all the new replacements since nobody wanted to do it.

Why? Because the old style radio weighed about 40 pounds and you already carried lots of weight. This was in February in a cold country. The other extra weight was of course more clothing, including the heavy rubber Mickey Mouse boots, your weapon, clips of ammo, hand grenades, and a flak jacket.

Most people felt that the radio man would be shot first by the enemy. On one of these patrols, I was walking in the footsteps in the snow of four or five other members who were ahead of me. Of course, each person sunk down in the snow more and more. I sunk way down and got stuck in the snow so deep, a member of the patrol had to come back and help pull me out.

One of the scariest jobs we had to do was go out at night, man the listening post, and stay until daylight. Two of us went out each time to a small fighting hole out in front of the lines. It was about 100 yards or more down on an L finger, or ridge of land.

We carried an Army field telephone and hooked it up to the wires that were previously placed there by others. We had to call back to our Platoon Bunker every hour to let them know everything was all right. We could call in anytime the enemy was near us.

We also carried to the listening post an M-1 Carbine rifle with a snooper scope attached and a battery pack. This snooper scope allowed us to see in the dark, although it was spooky to look through it. It was scary with just two of us out there. We were relieved when nothing happened and seeing daylight meant we could go back up to the line.

One time I went out to the listening post with Ray Fields, the full-blooded Pawnee Indian from my squad. The Chinese attacked before day break near our platoon area. Ray Fields and I were to the left of our Company area.

We could hear all the shooting and the explosion of rounds over on our right flank. We called on the telephone, and were told to stay put until daylight. We thought surely the Chinese would come up the finger where we were. But we lucked out as they did not come up. In fact, one morning a Chinese soldier came up to the bunker next to ours and surrendered.

When we got back to our Platoon area, we found that three or four members of our Platoon were wounded. One guy was from Iowa and was about to rotate home. Another was from my squad and he was sent to the hospital in Japan. Later on he returned to our squad.

It was here that we got rid of SFC Bench who was our Platoon Sgt. He was the 1st Platoon Sgt, of Company "C" and was a total jerk. The men in my Platoon, who had been around Sgt. Bench for a long time, told of many things that had happened and how poor he was as a Sergeant. After I was placed under his command, I can recall five different incidents where he was a jerk.

One day Sgt O'Reily, the 2nd squad leader who had been there the longest time, came around to every one in the platoon with a petition for us to sign. He was going to take it to our Company Commander because we hadn't had a Lt. assigned to us as Platoon Leader for some time.

The petition said something like this: *"We, the members of 1st Platoon, "C" Company want you to transfer one SFC Bench out of our Platoon for various reasons, or one or more of us will take care of him."* In other words, he would be killed. In less than an hour he was transferred out of our Company.

We went back down into Reserve Position again, where in the daytime we had to do dismounted drills, combat training, and other details. We also got to go to a nearby river where we could bathe. At night they showed old movies outdoors. We had a couple of company beer parties. We also had to guard an air base at night. This small air base had the small Piper Cub type planes that carried observers for the Artillery.

Wouldn't you know it; a member of our Platoon was over at this air base and saw SFC Bench. He was an artillery observer flying in one of those planes. The jerks get all the breaks.

One night after a company beer party a bunch of us left the mess hail and went into the cook's tent. One guy from my platoon was playing his guitar and singing while others joined him on some songs. Some of the guys were from another platoon. One of them was Teraholt, who we called the "Bad Bastard from Boston." He surely could sing the song "Danny Boy."

Teraholt got sleepy, so he wandered over to an empty cook's cot to lie down. It was quite dark in the tent since we had only one small candle for light. The tents were quite large. There was enough room for two squads of 20 men. It had 20 cots, with lots of room between the cots and an aisle down the middle, with two large tent poles a little ways in from the front and back tent flaps, that we called doors.

After his duty ended, the cook came into the tent. He went to his cot and yelled out, "Who the hell is laying on my cot?" Teraholt would not wake up. So four of the guys from his platoon went over and picked him up. They were going to carry him back over to his tent. They carried him out feet first with each of the two guys in front holding one of his legs. Remember it was quite dark in the tent. They were moving right along when Teraholt's legs straddled one of the large tent poles.

Need I tell you that Teraholt woke up abruptly? He was moaning and yelling, "I'll get you bastards for this in the morning." I am sure that nobody in his tent knew about this when asked the next morning. We never heard any more about it.

A few days later, during the night, they loaded us onto trucks, with full gear, and we were driven back to the front lines in the Mun Dung Ni Valley. We unloaded about three miles behind the front line, and walked uphill the rest of the way.

Someone from each squad of the regiment whom we were relieving would escort us up to the line to take over their positions. Then they would all quietly move out, and go back down the hill to the trucks we had come up in.

The pressure is greater each time you have to go back into combat, each time you go up to the front line. It is especially hard for those who have been there before. When it is your first time you do not think about it as much. If you were there before you wonder if you will be lucky again, or if you will be carried back this time. You are as scared as the new replacement next to you, but you do not show it.

When we went out on patrol one night, I was in charge of the last five or six members of the patrol. It was very dark that night. We would move a little ways, then stop and crouch down and listen for a while, then repeat this over and over until we reached our destination. We would stay far enough away from each other that we could just see the one in front of us.

Most patrols at night were ambush patrols. We would go out and set up in a defensive circle and attack the enemy if they came near us. That night the guy in front of me was Amy, a short black soldier from my squad. I saw him stop so I stopped and all the rest behind me stopped.

I waited and waited. I could still make out the outline of Amy crouching ahead of me. We sat there for some time. I finally went back and told those behind me that I was going up to find out why we were not moving.

I moved up to where Amy was supposed to be, and he was not there. There was only a small evergreen bush about the same size as Amy when he crouched down. We were out there in a lot of small hills and open fields, and Amy had to stop right by the only evergreen around.

The patrol was long gone ahead of us. Where, I did not know. I went back and told those behind me what had happened. I told them I did not know where the patrol was, but I was going to try and find them. I took off and a while later I found them. To this day, I do not know how I did it. It must have been from experience. Our patrol in front never knew what happened to us. It was just another, "what if" situation.

While we on the front lines at the Mun Dung Ni Valley, there was a tank on our right flank dug into a permanent position on the line facing the Chinese. Every now and then those 'damn tankers' would fire off a couple of rounds at the Chinese. We hated that. When they did it we would cuss them out, because it was not long after that the Chinese would start firing mortars and artillery shells at us. We had some close calls.

Soon we were pulled off the front lines, put on trucks, and moved out to a town on the eastern coast of Korea. The whole regiment plus the tank unit attached to us were loaded onto LST's. We went south and west of Korea to Cheju Do Island. We relieved a regiment from another division, and took their place guarding Chinese Prisoners of War.

We were there for a month or so before we were relieved by another regiment from another division. They put us back onto LST's for the return trip. Then we were loaded onto trucks that took us about 20 miles behind the front lines. They put us in a forced march up a road for 10 miles and we all had blistered feet when we stopped. We were spread out over a few small hills where we put up our pup tents.

A few days later they issued us new tents that were made in a large circle that would hold from four to six men. We went to sleep one night and when we got up, we had five inches of snow. We stayed there until about the middle of December, 1952 when we were told to get all our gear together and make full packs. We were on our way back up to the front lines in the Kumwha Valley area.

As I said before, each time you go back up to the front lines, it gets to you. You may not say anything to anybody, but you sure do to yourself. I had been up there three times and was on my way again.

We had marched along this road for about three miles when a Jeep pulled up and a soldier yelled, "Wilson, get in the Jeep. You are going home." That was music to my ears. I was really relieved this time. My only regret is that I did not get the home addresses of many of my fellow soldiers.

Cpl Robert C. Wilson 3rd Squad — 1st Platoon,
Charlie Company, 35th Infantry Regiment, 25th Infantry Div.

The following article first appeared in the Spring 2002 Edition of *Tropic Lightning Flashes* and is reprinted by permission of The 25th Infantry Division Association.

MONSOON SEASON IN KOREA by Cpl Robert C. Wilson

In every period of War there are Service Men and Women that are injured or killed in Basic Training, and in combat zones, by mistakes, accidents, miss-communication, frightened individuals, previous placement of mines and booby traps by our own troops, short Artillery rounds, bombs dropped in the wrong place, your own troops shooting at you, when returning from a Patrol, and by natural disasters such as hurricanes, fires, floods and cave ins.

I was a BAR man (Browning Automatic Rifleman) on the front lines, in the Mundung Ni Valley, near Heart Break Ridge. In July it had been raining 24 hours a day for many days. There were 10 men in my squad, who were split up into three Bunkers. There were three in my Bunker; including an older soldier from Oklahoma with the last name Maxwell. He was recently assigned to our squad. The other two were Joseph O'Brien, "OBE," our squad leader from Georgia and myself from Flint, Michigan.

On July 30, 1952 I was pulling my shift of guard duty, out in a hole in front of the Bunker. It was late and very dark. At this time I was in the 3rd squad, 1st Platoon, Company C., 35th Infantry Regiment 25th Infantry Division. After I had pulled my time out in the hole, I went back inside the Bunker to wake up Maxwell to take his turn out in the hole. Then I went back out there and waited for him to come and relieve me.

I went back into the Bunker to get some more sleep. My bunk was just inside the poncho covered doorway, to the left and against the back wall of the Bunker. My bunk was made high off the ground, framed out of old small trees and wrapped with some chicken wire. Our Bunker had a larger room connected to the rear of the small room where I slept. Sgt O'Brien slept on the top bunk in the larger room; and Maxwell had the lower bunk.

I sat on my bunk and took off my boots. I was about to lie down when I heard a noise behind me. I immediately got down and went in to the other room and woke up O'Brien and told maybe we should leave the bunker, as it might collapse. He got right up and we went back to the front of my bunk. I sat up again on my bunk; when again I heard the noise of dirt moving. I looked behind me and I could see openings between the logs. I jumped off of my bunk and pushed 0'Brien out the door, but I was not fast enough to make it all the way out myself.

The bunker came down on top of me, covering all of me except my head, and right arm. Our bunker was made of logs, surrounded by sand bags. The top of it had a lot of weight. There were logs, layers of sand bags, dirt, and rocks, and it was all water logged. All of this weight was on top of me but the worst part was that somehow a log was under my stomach, and I was being squeezed by all that weight on top of me. I remember that I shook hands with 0'Brien and told him goodbye, as I could hardly breathe.

Lucky for me in that pitch black wet night, "OBE," Maxwell, and others who heard what happened, came running over to help get me out. I found out later that some of those men broke their finger nails off getting me out of there. I can still to this day feel the relief I felt after they lifted most of the weight off my back. While this was going on, someone called down the hill to the rear and got the Battalion Aid Station to send up two medics in a litter Jeep to get me.

There was a narrow road that came up the hill behind our lines. It went past our Company C, past Co B, on our right flank a ways, where there was a turn-around. The road was hub deep in mud when the Medics came up to get me; in fact they could not make it all the way up due to the mud; so they carried me down to the Jeep on a stretcher.

I think the Medics gave me a shot of morphine, as I was in pain, and I was unconscious at times. We started down the hill in the Jeep and we suddenly stopped. The two Medics jumped out of the Jeep. One of them was franticly chopping away at a tree that had fallen across the road after they had passed on their way up the hill, while the other Medic was trying to console me. They took turns doing this until they could remove the tree.

We finally got down to the Battalion Aid Station, where I was checked out by a doctor. I was still going in and out of consciousness.

I think they gave me another shot, and I remember the doctor saying, "We will keep him here tonight, then send him down to a MASH unit in the morning." Next morning I could hear them debating whether to call for a helicopter, or to use an Army ambulance to take me back to the rear. They decided to use the ambulance, and just a driver and I left the Battalion Aid Station.

On the way there we were crossing a stream when the engine quit, right in the middle of the stream. I felt sorry for the driver, a young Soldier. He did not know what he should do. He opened the back door, and was kind of talking to me, but mostly to himself. He said, "What should I do? Should I take you out of the ambulance, and carry you to the bank? Should I just wait here hoping some one will come by? Should I run down the road to find some one who can help?"

I guess he ran down the road. It was quiet for awhile and then I heard people talking. Then the ambulance was pulled out of the water and they got it started. We were on our way and a short time later we were at the Hospital. They took off all my clothing, (my boots were still up in the fallen Bunker). They put salve on my skin where it was scraped off from my shoulders to the back of my legs and a portion of my right heel. I lay on an Army cot between sheets for eleven days. After 11 days I was issued clothes, and they took me back up to my Regimental Command Area, where I spent the night.

The next day I was back up on the front lines with my squad. They had rebuilt a new bunker while I was gone. Some time later we were pulled off the front lines and put on LST's where we were shipped south to the Korean Island Che Ju Do. We were to guard Chinese Prisoners of War. While there I was sent to Japan for R and R. which was supposed to be for a total of 10 days from my Company. But due to Army error, I was gone for 25 days.

When I got back to the Island I found out that Joe O'Brien had shipped out, back to the States. I was very disappointed, as I did not get to say good bye. Even more important I did not get his home address. "OBE" never talked about his home life and he never wrote home. In fact, one day while up on the front lines there was a visit from someone from the Red Cross who wanted to know why he was not writing home to his Mother. All I can remember is that he enlisted in the Army to make it a career. Before coming to Korea, he was in the Airborne in Japan with the 187th Airborne Regiment, and he was originally from Bismark, Georgia. I tried to locate him using a computer, but no luck. I know in my heart that I owe my life to Joseph O'Brien and the others who dug me out that wet dark night in Korea.

Sgt Robert C. Wilson, US Army,
Co. C. 1st Bn, 35th Infantry Regiment

Left: Wilson standing outside the entrance to the bunker that almost killed him. Right: Joseph "Obe" O'Brien and Wilson outside the bunker before the fateful collapse.

CUMMINS BROTHERS
Lewistown, OH

This is a heartfelt tribute to the author's three brothers, who served during the Korean War.

"I'm sorry for not listening to you and remembering your stories."

—William A. Cummins

Photos were taken in 1951 on our farm in Lewistown, Ohio.

Left: Dick. Right: From left Vern, Bob, Bill, and Jack.

- Richard E. Cummins (Died 2006)
- PFC LAVERN R. CUMMINS, US ARMY (Died 2007)
- A1C ROBERT L. CUMMINS, USAF
- William A. Cummins, Author of THE FORGOTTEN
- SGT JACKSON F. CUMMINS, USAF (Died 2008) — Stayed 20 years in the USN and USAF — Served aboard Navy Destroyer patrolling the Korean shoreline in 1952.

WRAP UP

THIS BOOK IS DEDICATED TO ALL OUR UNSUNG HEROES. TOO many Veterans have already passed from this world with their stories still untold. CAI Publishing wishes to thank all the Veterans who sent their stories. Those not selected for the first volume of *THE FORGOTTEN* will be published in future volumes. Our belief is that all Veterans who wish to share their personal stories in a book should be able to do so. We are proud to offer this venue for that purpose.

THE FORGOTTEN VICTORY

A Country's Salvation

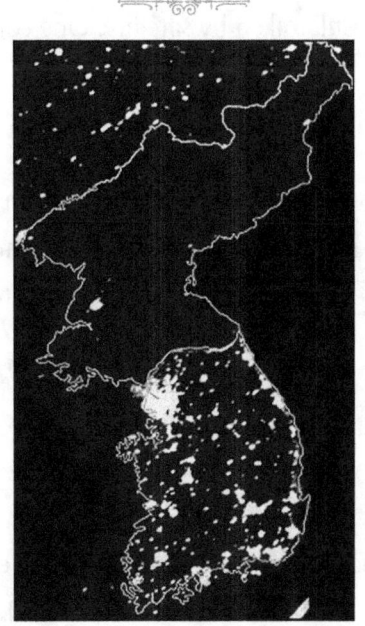

Look closely at the photograph.

Why Is There So Much Light In South Korea
and
Only A Slight Glimmer In North Korea?

THE CONTRAST OF LIGHTED CITIES MAGNIFIES NOT ONLY THE LACK of electricity, but lack of business activity in the North Korean cities. Nothing sums up the "Forgotten Victory" as well as this photograph of the Korean Peninsula taken by satellite. One country and one people separated by the DMZ in 1953.

How Can this Be?

South Korea has come into the 21st century as an advanced country while North Korea still has the designation of third world country. Are the people different from each other? Absolutely not! They are in fact related with families on both sides of the DMZ. How did South Korea develop into a bustling civilian and commercial environment that now stands head and shoulders above its northern counterpart?

Let us Take a Look...

The information that follows may help explain the progress and the rapid development of industrialization by South Korea as well as the stifling of such growth in North Korea.

Chapter One

KOREAN WAR — GLOBAL PERSPECTIVE

"All wars are waged by Good and Evil for domination of the human spirit."

—William A. Cummins

SERIOUS REFLECTION ON THE KOREAN WAR ENTAILS THIS VERY important observation. This war was intentionally waged for domination of the human spirit. It was not caused by a disagreement between like-minded countrymen.

During WWII, the United States, China, and Russia were united against Germany and Japan. However, all the nations knew the United States was fighting for human freedom, not for another form of human tyranny. The biggest problem for the Communists after WWII was how to deal with that freedom loving peaceful country, The United States of America.

To test America's resolve, Communist troops of North Korea invaded the democratic Republic of South Korea on June 25, 1950, leading to the bloody Korean War. It soon became apparent that their superior military forces would easily take over the entire Korean peninsula. It was feared this tiny peninsula would erupt into World War III.

When the Soviets decided to boycott the United Nations and not exercise their veto power, the United Nations was able to intervene by joining forces with the United States and the Republic of South Korea to stop the invasion.

Twenty-two nations responded to the UN's call to defend freedom and repel the Communist aggression. If not for swift actions by the free world countries, the Communist's plan may have worked. North Korea not only had the military support of China but the support of the Soviet Union government. The stage was set for a bloody three year war.

The two armies crisscrossed the 38th parallel dividing line several times. When the Chinese feared that their own borders were threatened, they became involved on the side of the North Koreans. The conflict then escalated greatly.

During the war several decisions were made that set the course of World history. Prior to the Korean conflict, America was busy disarming from WWII and ignored the threat of the Communists. Following the North Korean invasion President Truman established the doctrine that no country would fall to Communism. This decision marked the beginning of the end of the Soviet Union and established our industrial base for at least 50 years.

By 1953, the Korean War was actually at a stalemate and technically, it still is. The US Congress never declared war and the July 27, 1953 truce that ended it was not a peace treaty, only a ceasefire. There never was a military victory for either side. The Americans, who went to the polls in late 1952, just wanted the war to end and were willing to accept a ceasefire outcome.

If the Korean War was not a military victory, then where was the victory? It has been 56 years since the war ended, and the world still faces the aggressive Communist regime in North Korea. Even now, the United States, South Korea, Japan, China, and Russia are all focused on one of today's most important goals which is stopping North Korea's nuclear weapons program.

As we look back from the 21st century, the historical significance of the Korean War becomes apparent. At that time many of the world leaders were seriously debating the merits of Communisam vs. Capitalism. Those defenders of freedom for South Korea actually paved the way for the advancement of Capitalism around the globe.

From all appearances, this lonely forgotten war actually stemmed the tide of global Communism. With this in mind let us now take a look at the forgotten war that resulted in a forgotten victory for South Korea and the free world.

Chapter Two

KOREAN WAR — HISTORICAL VIEW

KOREA IS A NATION ABOUT THE SIZE OF THE STATE OF UTAH. IT IS an ancient civilization located on a peninsula jutting southward from the east coast of Asia where it developed from walled-town states and larger kingdoms.

Korea was united in the 7th Century and was opened to trade by Japan in 1876. Both China and Russia competed for influence until the Japanese annexed the country in 1910. The Korean peninsula under Japanese rule refers to the period between 1910 and 1945 when Korea was forcibly annexed by the Japanese Empire.

Japanese control of Korea ended with the surrender of Japan to the Allied forces in 1945 at the end of World War II. The Korean Peninsula was subsequently divided into South Korea and North Korea along the 38th parallel. The legacy of Japanese occupation and the atrocities committed against the Koreans by the Japanese has caused continuing disputes between Japan and the two Koreas.

In a proposal that was opposed by nearly all the Koreans, the United States and the Soviet Union agreed to occupy the country as a temporary trusteeship with the zone of control demarcated along the 38th parallel. This type of arrangement acquired semi-permanent status with the onset of the Cold War between the free world and Communism.

In August of 1945, the Soviet Army established a Soviet Civil Authority to rule North Korea until a domestic regime, friendly to the USSR, could be established. On September 7, 1945, the US Army placed South Korea under the direct administration of US military units with Lieutenant General John R. Hodge as the administrator.

Three years later, on August 15, 1948, South Korea founded the Republic of Korea (ROK). Less than one month later, on September 9th, 1948, The Democratic People's Republic of Korea (DPRK) was founded in North Korea.

The political leaders of the free world wanted to believe WWII had brought about a peaceful era around the globe. But the leaders of Russia and China decided it was an ideal time to test the willingness of the United States to fight another war for human freedom. They thought conditions were ripe for Communism to rear its ugly head and continue its spread of personal destruction across the globe.

They decided to test America's resolve in Korea, where they had successfully divided both the country and its people after WWII. China believed it could defend its bordering neighbor, North Korea, should the United States intervene, so they crafted a plan to defy the world. They would quickly clear the southern peninsula of any South

Korean military resistance and consequently drive the American armed protectors into the sea.

Thus the Korean War began in the early morning hours of June 25, 1950, when the Communist army of North Korea launched an attack into South Korea. The North Korean offensive started from four locations across the 38th parallel into South Korea.

In only 41 days, the South Korean and American forces were driven south into the Pusan perimeter, just a few miles from the southern shore of the tip of South Korea. By August of 1950, the South Korean Army collapsed and it looked as if they would lose everything.

President Truman decided the US was not going to abandon Korea. A United Nations Command (UNC) was quickly established to assist South Korea under UN Security Council resolutions. More American troops were committed and after some tough times, the conflict stabilized around the Pusan perimeter.

It stayed like that until mid-September when MacArthur landed US troops at Inchon. He directed a breakout of the Pusan perimeter the next day and began an operation that cut off the major portion of the North Korean Army. Tens of thousands of prisoners were taken, and within 10 days the war started to disintegrate.

After a successful perimeter defense near the southeast city of Pusan and US landings at Inchon in September, the UN and South Korean forces drove the North Korean forces back north and Seoul, the capital of South Korea, was recaptured and liberated. On September 29, MacArthur and the Republic of Korea (ROK) President, Syngman Rhee, held a victory parade in Seoul. This event was hailed as the end of the war.

Meanwhile, Communist China warned the US through its envoy in Switzerland and other channels, that it would fight if the UN forces crossed the 38th parallel. MacArthur ignored the Chinese rhetoric and the UN allied command changed the rules. Instead of simply getting the original territory back, the UN seized the opportunity to unify all of Korea and decided to press on toward the Yalu River and the Chinese border.

Three months later, the Marines and forward details from the Army, together with British, French, Turkish, South Korean and other United Nations forces stood at the Yalu River, the border between North Korea and China, thinking the war was nearly over.

At about the time the UN forces reached the Yalu River, China entered the war. While the UN forces were advancing north, a force of 300,000 Chinese troops had moved south into North Korea, and concealed their troops in the mountainous terrain. On November 25, 1950, they attacked the UN forces from the rear, and "all hell broke loose."

With hordes of Chinese forces joining the war, the UN forces had to fight their way back south to the east coast of North Korea, where they were rescued by the Navy and taken to more secure positions in South Korea.

The battle line was again pushed south of Seoul before the UN and South Korean forces held them. During the next two and a half years, the Korean conflict became trench warfare and battles for hilltops were fought back and forth across the 38th parallel.

The war devastated the entire Korean peninsula. Seoul changed hands four times and was reduced to rubble. With its air supremacy, the UN Command was able to destroy almost every building of importance in North Korea.

Over one million died on each side, but North Korean losses were the greater. An armistice was signed between the UN Command and North Korea/China on July 27, 1953. South Korea refused to sign the armistice, but agreed to abide by its terms.

The July, 1953 armistice split the peninsula along the demilitarized zone near the original demarcation line. Since no peace treaty was ever signed, the two countries are still technically at war, and are only separated by a Demilitarized Zone called the DMZ.

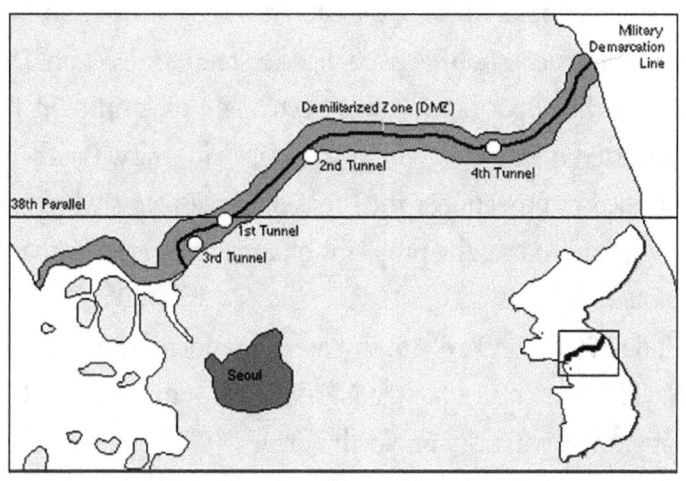

The Korean War Demilitarized Zone known as the DMZ

For the three ensuing decades, the Republic Of Korea in the south was led by conservative presidential dictators who restricted most political rights because they were so vehemently anti-Communist.

The first President, Syngman Rhee, made full use of his authority. He revised the Constitution and stayed in power until forced to step down after a student uprising in 1960.

General Park Chung-hee staged a military coup in 1961 and remained in office until assassinated in October 1979. Then another general, Chun Doo-hwan, assumed power the following year, after martial law was declared.

Park's policies did little for political development but they encouraged economic development through promoting business conglomerates. The rapid economic growth raised people's expectations and led to the development of a democratization movement that could no longer be ignored by the government in the mid-1980s.

Faced with this and increased international attention leading up to the Seoul Olympics in 1988, the chosen successor to Chun Doo-hwan, former general Roh Tae-woo, had to stand for election. Due to splits within the opposition, he was elected and took office in 1988.

This started the development of a strong democracy. The economy has continued to grow under the succeeding presidents who have increasingly had to face the problems of a maturing, rather than a developing, economy.

In June 2000, as part of South Korean president Kim Dae-jung's Sunshine Policy of engagement, a North-South summit took place in Pyongyang, the capital city of North Korea.

That year, Former President Kim received the Nobel Peace Prize "for his work for democracy and human rights in South Korea and in East Asia in general, and for peace and reconciliation with North Korea in particular."

In 2002, South Korea and Japan jointly co-hosted the 2002 FIFA World Cup Soccer finals. The event marked South Korea's emergence onto the world stage and provided even stronger economic growth and a cultural union between South Koreans.

The South Korean national football team became the first and only Asian team to reach the semi-finals, beating Spain, Portugal and Italy in the knock-out stages.

In 2005, South Korea's economy broke the US $1 trillion mark. Its per capita GDP is now comparable to that of Italy, Greece, New Zealand, Portugal and Spain. Its economy is continuing to grow rapidly and is forecast to surpass that of Canada, France, Germany and the United Kingdom by 2025, South Korea's per capita GDP is then expected to overtake that of Japan by 2050.

Chapter Three

KOREA TODAY — THE VICTORY

A COMPLETE STUDY OF KOREA IS THE WORK OF A LIFETIME. Therefore, only short introductions to a few facets of South and North Korea are presented here.

Korea, a nation about the size of the state of Utah, is located on a peninsula jutting southward from the east coast of Asia. This map shows some of the principal cities, rivers, and mountains of Korea. The land is inhabited by the Korean People, almost unmixed with any other ethnic groups.

A country long unified and peaceful, Korea today is a divided land. Its people are split between the Communist North and the capitalist South. Even the flags of the respective countries show their differences. The flag of the North features the Red Star of Communism, while the flag of the South features ancient philosophical symbols. The circular "TaeGuk" symbolizes the harmonious state of yin and yang, and the four "Kwe" symbols represent Heaven, Fire, Water and Earth.

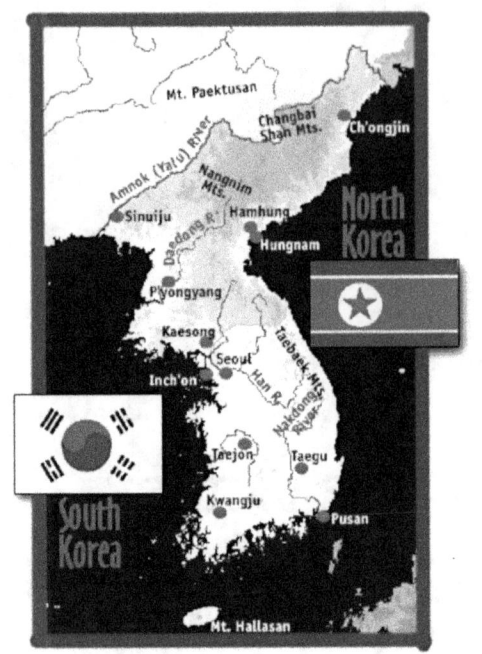

Korea is a beautiful land largely covered by mountains. The main mountain range, the Taebaek covers most of North Korea and extends south-southeast through the east center of the peninsula. The mountains are cut by many narrow river valleys, but the only wide valleys and plains are along the west coast.

The climate of Korea is not unlike that of the eastern United States, with hot wet summers and cold dry winters. The combination of elevated topography and high rainfall in the North provides for ample hydroelectric power.

Korea has one of the highest average population densities in the world, comparable to the most heavily populated areas in the United States and Europe. The mountain highlands in the northeast and south are generally less populated than the coastal plains in the west.

One of the most important population phenomena in Korea since the Korean War is the shift of population from rural to urban settings. More than three-fourths of all Koreans lived in the countryside in 1950, but now more than two-thirds of the North Koreans, and more than four-fifths of the South Koreans live in cities.

Even though the population of both nations has more than doubled since the Korean War, the rural population of North Korea has stayed about the same, while the rural population of South Korea has decreased by more than half.

The rate of growth in the last ten years has slowed greatly due to government efforts. The birth rate in Korea has now dropped below replacement level. As a result, the average age of the general population will grow older in the future.

Due to population shifts, city populations in Korea have exploded in the last 15 years. Seoul has led them all. A city of only about 1 million people in 1950, it now has over 11 million people. Seoul is currently the tenth largest metropolitan area in the world, comparable to Mexico City, New York City, Los Angeles, and Bombay.

The migration to the cities has been fueled by the lack of land and jobs in the countryside. Life in the city may not be better than in the country, but it certainly is different. Working conditions range from hard physical labor in trades like construction and assembly plants, to typical desk jobs in modern business buildings and school campuses. Jobs are plentiful, but living space is limited and expensive.

Conditions in the large Korean cities are similar to those in large American cities with smog, traffic, freeways, and big and sometimes ugly buildings. There may be more bicycles than in a typical US city, and the signs are in a foreign language, but otherwise a snapshot photo could be in any large city in America.

They are all Koreans — there are no other ethnic groups present. The populations of both North and South Korea are almost perfectly homogeneous. The last census in South Korea indicated only one other ethnic group of about 20,000 Chinese.

Reminders of Korea's past are found everywhere in both North and South Korea. The Korean people feel strong ties to their past and to their cultural heritage. Traditional arts of painting, drama, and ceramics are kept alive by both individuals and government support.

Religion has been an important part of Korean life for more than a millennium. Religious activity is discouraged under the Communists in North Korea, so about half of the people there are nonreligious. On the other hand, in South Korea religious thought and activity permeates much of their society. The most prominent religions in South Korea are Buddhism, Christianity, Confucianism, and a few local religions that combine aspects of all three.

The economy of a nation depends on the characteristics of its people and available natural resources. Korea as a unified nation was blessed with an industrious people and abundant natural resources, though the natural resources were not uniformly distributed. At the time of the partition, North Korea had most of the minerals, energy resources, and heavy industry, while South Korea had most of the light industry and agricultural lands.

Most of the country's industrial and energy facilities were destroyed in the Korean War. In the mid-1950's, both North and South Korea essentially started rebuilding their respective economies from the ground up.

The economies in the two Koreas have been rebuilt during the last 40 years according to the different economic philosophies of their respective governments and within different economic communities.

Those different economic philosophies are also world views dividing humanity into captivity under Communisam and freedom under Capitalism. The nations of North and South Korea represent both sides of the same coin and reveal how the same ethnic culture must adapt in order to co-exist under different governments.

NORTH KOREA — A NUCLEAR THREAT

North Korea is the last Stalinist state on earth. Under Communist rule since 1945, its international relations have shrunk considerably. Tied by politics and ideology to the Communist bloc after WWII, it has developed few ties to the rest of the world.

The core ideological objective of North Korea leaders is eventual unification of both nations under Communist rule. Through state-funded propaganda, political, economic, and military policies North Korea demonizes the United States as the ultimate threat to its social system.

North Korea's history of regional military provocations, the proliferation of military-related items, long-range missile development, and weapons of mass destruction programs are of major concern to the international community.

In recent years, North Korea has been viewed as a growing threat to peace in the rest of the world. It has established and armed a huge standing military force of over a million soldiers. Annual defense outlays are estimated to be about $5 billion, an astounding 20 to 25% of its GDP.

After decades of economic waste and mismanagement North Korea has relied heavily on international aid to feed its population.

North Korea has also developed its own short-and intermediate-range guided missiles. Although the missiles are primarily intended for use against South Korea, they are apparently being sold to other nations like Libya.

The New York Times reported in January of 2009 that North Korea's leader has declared an "all-out confrontational posture," stating they had "weaponised" enough plutonium for roughly four to six nuclear bombs.

After receiving the threats, South Korea ordered its military to heighten the vigilance along its heavily fortified border with North Korea.

Amidst concern about North Korea's nuclear weapons program a series of six-party talks has been established with diplomats from the US, China, Japan, Russia, South Korea, and North Korea meeting to discuss elimination of nuclear weapons from the Korean peninsula.

For the sake of world peace and stability, the United States continues to work with its six-party partners to move forward with the denuclearization of North Korea.

SOUTH KOREA — THE ASIAN TIGER

In contrast to North Korea's failures, South Korea is prospering. Known as the "Land of the Morning Calm," it is neighbored by China to the west, Japan to the east and borders North Korea to the north. South Korea's capital and largest city is Seoul, the second largest metropolitan city in the world.

This Asian Tiger is among the worlds fastest growing and developing countries. Today, its success story is known as the "Miracle on the Han River," a role model for many other developing countries.

South Korea is leading several key industries in the world, particularly in the fields of science and technology. It has a very advanced and modern infrastructure and is a world leader in information technology such as electronics, semiconductors, LCD displays, mobile phones, computers, and others led by Samsung and LG.

Home of the world's third largest steel producer, POSCO, South Korea is the world's largest shipbuilder, the world's fourth largest oil refiner and one of the world's top five automobile producers, headed by Hyundai and Kia. It is also a leading country in biotechnology, robotics, textiles, construction, engineering, petrochemicals, and machinery.

South Korea's geopolitical realities during the Cold War era caused South Korea's leadership to maintain close political and economic ties to the United States and the West after the end of WWII and the Korean War.

Even during the 1960s and '70s, when the government was essentially a military dictatorship, ties with the West remained strong. With the establishment of a civilian government in the late 1980's, their international ties only became stronger. Broader International cooperation is a major diplomatic goal of South Korea. In 1992, it established diplomatic relations with mainland China.

South Korea's most notable exception to good relations is neighboring North Korea. There have been periods of political "thaw" between the two nations and sentiment for rapprochement is growing in South Korea.

In June 2000, as part of South Korean president Kim Dae-jung's Sunshine Policy of engagement, a North-South summit took place in Pyongyang, the capital city of North Korea. That year, Former President Kim received the Nobel Peace Prize "for his work for democracy and human rights in South Korea and in East Asia in general, and for peace and reconciliation with North Korea in particular.

However, international relations remain unsteady, and fear of invasion by the North is constant. Negative attitudes toward the North are still quite prevalent in large segments of South Korea's society.

In order to protect itself from the northern threat, South Korea maintains an armed force of over 600,000 military personnel, distributed throughout its Army, Navy, and Air Force. Equipment and training are kept to the highest modern standards.

South Korea is a major non-NATO ally with the world's sixth largest armed forces, the world's second largest force of reserve troops, and the tenth largest defense budget in the world. All South Korean males are constitutionally required to serve in the military, typically for a period of two years.

The South Korean Army has 2,300 tanks in operation consisting of technologically advanced models such as the K1A1 and the new K2 Black Panther. The South Korean Air Force operates the ninth largest air force in the world.

The South Korean Navy has the world's twenty-sixth largest fleet of destroyers. It is one of the five Navies in the world to operate an Aegis guided missile enabled destroyer. It has the world's largest fleet of frigates, the sixth largest of corvettes, and the fourth largest of submarines in operation.

From time to time, South Korea has sent its troops overseas to assist American forces. It has participated in most of the major conflicts that the United States has been involved in during the past 50 years.

GENERAL INFORMATION ON SOUTH KOREA

Official Name: Republic of Korea (ROK)

Area: 38,022 square miles — about the size of Indiana — covering approximately 45% of the Korean peninsula.

Population: 49.05 million

Capital City: Seoul with a population over 10 million.

People: Korean with a small Chinese minority

Language: Korean. English is widely taught in schools

Currency: ROK Won (KRW)

GDP Per Capita: $24,500

Religion(s): Wide range from Shamanism, the oldest, to Buddhism, Confucianism, Chondogyo, and Christianity.

Major Political Parties: Grand National Party (GNP); United New Democratic Party; Liberty Forward Party (LFP); Pro-Park Alliance (PPA); Democratic Labour Party (DLP).

Government Type: Republic — A presidential system backed by a unicameral National Assembly of 299 members.

President: Lee Myung-bak — elected December 2007.

SOUTH KOREA HONORS OUR VETERANS

My name is Corporal Robert McGuire, US Army (Ret) Daytona Beach, Florida. During my trip to Korea, in June 2008, the following message resounded from all quarters:

> "The Korean War Is Not Forgotten In Korea.
> Those Who Came To Save Our Country
> Will Never Be Forgotten Here."

While serving in the US Army during the Korean War, I observed the resourcefulness of the Korean people. Their skills became apparent during the stress, strain, dislocation, and death of millions during the fighting.

Suddenly left without buildings and shelter, they used the wooden equipment pallets discarded by UN Forces to fashion homes and shelters of all types. In some places they even made store fronts from which to market their wares that they gleaned from the American trash piles.

Expended shell casings were turned into belt buckles, ash trays, lamps and jewelry, while beer cans were turned into useful items for the GI's who were weary from fighting. Combining resourcefulness with patience, the Koreans have shown the world what they can do when given the opportunity.

I joined with many veterans for a trip called the 'Revisit Korea Program' sponsored by the US Korean War Veterans Association (KWVA) and the Sae Eden Presbyterian Church is located in Seoul, Korea.

When we landed in Seoul, for as far as the eye could see, the city that had been reduced to rubble had risen from the ashes to be the envy of many nations on earth. It now ranks as one of the strongest economies in the world.

During our revisit, we were treated like returning kings and queens. Children in Korean attire were waving flags and chanting, "You are my hero" as they greeted us at the entry to the church. We were then given a standing ovation by the 5,000 parishioners awaiting our arrival.

I was 18 years old when I first arrived in Inchon harbor in 1952 aboard a landing craft. On this trip we arrived in a huge 747 and the greeting received brought tears to the eyes of every veteran. We suddenly had an overwhelming sense that our efforts and sacrifices on behalf of the South Korean people had not been in vain.

We finished our week by touring the Republic of Korea war sites, Panmunjom, 1st ROK Division Headquarters, The Korean War Memorial, and ancient cemeteries along with numerous celebrations and banquets.

It ended with a 58th War Anniversary Luncheon hosted by the Korean Government. Each veteran received a special medal with this Proclamation from His Excellency, Lee Myung-bak, President of the Republic of Korea:

> "It is our great honor and pleasure to express the everlasting gratitude of the Republic of Korea and our people for the service you and your countrymen have performed in restoring and preserving our freedom and democracy.
> "We cherish in our hearts the memory of your boundless sacrifices in helping us reestablish our Free Nation.

"In grateful recognition of your dedicated contributions, it is our privilege to proclaim you an "AMBASSADOR FOR PEACE" with every good wish of the people of the Republic of Korea.

"Let each of us reaffirm our mutual respect and friendship that they may endure for generations to come."

The President then gave this final declaration to every representative of the 22 UN countries in attendance: "Go home and tell everyone you know, '***This is your homeland too!***'"

In summary, if war has any lasting value it can only be measured by the results. The Korean War, by any measure, was a Victory for Freedom!

Respectfully,
Cpl. Robert McGuire
US Army (Ret)

Chapter Four

KOREAN WAR VETERANS MEMORIAL

"Freedom Is Not Free."

These four words on the wall of the Korean War Memorial reflect the sentiment of men and women who served in the Korean War—as well as those who fought and sacrificed to preserve democracy throughout our Nation's history.

FROM 1950 TO 1953, THE UNITED STATES JOINED WITH UNITED Nations forces in Korea to take a stand against what was deemed a threat to democratic nations worldwide. At war's end, a million and a half American veterans returned to a peacetime world of families, homes, and jobs—and to a country long reluctant to view the Korean War as something to memorialize.

But to the men and women who served, the Korean War could never be a forgotten war. The passing of more than three decades brought a new perspective to the war and its aftermath. The time finally arrived, in the eyes of the Nation, to set aside a place of remembrance for the people who served in this hard-fought war half a world away.

On October 28, 1986, the US Congress authorized the American Battle Monuments Commission to establish a Korean War Veterans Memorial in Washington, D.C., to honor the members of the US armed forces who served in the Korean War.

Today the Memorial honors all our nation's sons and daughters who worked and fought under the most trying of circumstances, and especially those who died for the cause of freedom. It honors those who answered the call to defend a country they never knew and a people they never met.

One-and-a-half million American men and women, a true cross-section of the Nation's populace, struggled side by side during the conflict. They served as soldiers, clerks, chaplains, nurses, and in a host of other combat and support roles.

Many risked their lives in extraordinary acts of heroism. Of these, 131 received the Congressional Medal of Honor, the Nation's most esteemed tribute for combat bravery.

The memorial is designed as a place for reflection. The memorial is a circle intersected by a triangle. Visitors approaching the memorial come first to the triangular Field of Service. Here, a group of 19 stainless-steel statues, created by World War II veteran Frank Gaylord, depicts a squad on patrol and evokes the experience of American ground troops in Korea.

Strips of granite and scrubby juniper bushes suggest the rugged Korean terrain, while windblown ponchos recall the harsh weather. This symbolic patrol brings together members of the US Air Force, Army, Marines, and Navy; the men portrayed are from a variety of ethnic backgrounds.

A granite curb on the north side of the statues lists the 22 countries of the United Nations that sent troops or gave medical support in defense of South Korea. On the south side is a black granite wall. Its polished surface mirrors the statues, intermingling the reflected images with the faces etched into the granite. The etched mural is based on actual photograph of unidentified American Soldiers, Sailors, Airmen, and Marines. The faces represent all those who provided support for the ground troops. When viewed together these images reflect the determination of US forces and the countless ways in which Americans answered their country's call to duty.

The adjacent 'Pool of Remembrance,' encircled by a grove of trees, provides a quiet setting. Numbers of those killed, wounded, missing in action, and held prisoner-of-war are etched in stone nearby. Opposite this counting of the war's toll, another granite wall bears a message inlaid in silver which says, *Freedom Is Not Free.*

The Korean War Veterans Memorial Advisory Board was appointed by President Ronald Reagan to recommend a site and design, and to raise construction funds. Ground was broken in November, 1993. Frank Gaylord was chosen as the principal sculptor of the statues and Louis Nelson was selected to create the mural of etched faces on the wall.

On July 27, 1995, the 42nd anniversary of the armistice that ended the Korean War, the Memorial was dedicated by President William J. Clinton and Kim Young Sam, President of the Republic of Korea.

Everyone is invited to visit the Memorial which is staffed from 8 a.m. to midnight every day of the year, except December 25, by park rangers who are available to answer questions and give talks. A book store in the nearby Lincoln Memorial sells informational items relating to both the memorial and the Korean War.

The Korean War Veterans Memorial is one of more than 370 parks in the National Park System that represent our nation's natural and cultural heritage. Address inquiries to

*Superintendent, National Capital Parks Central,
900 Ohio Drive SW, Washington, DC 20024-2000*

APPENDIX

WANTED: MORE VETERAN STORIES

> "Something your children and grandchildren would like to see in a book."

TO: Veterans and Families of All Wars
FROM: William A. Cummins, Author

THIS IS THE FIRST VOLUME IN A SERIES OF BOOKS ENTITLED *THE FORGOTTEN* to honor our nation's war veterans. Please consider this an open invitation to all the veterans who would like to contribute their stories in their own words.

Write something your children and grandchildren would like to see in a book. Simply send three or four war related recollections. They can be serious, humorous, inspirational, or any combination, and may include photographs.

Include your current mailing address, phone number, and email address along with your rank and career path after leaving the military.

Veterans in this volume have been pleased with the way their stories were presented. I will be honored to have your story as part of the future work and welcome any assistance you can provide.

Stories and pictures can be slow mailed to the address below, or emailed to: wacummins@bellsouth.net. Please let me know if you have any questions. Phone: 386.383.5198.

Very respectfully,

William A. Cummins

William A. Cummins, *Author*
807 Black Duck Drive
Port Orange, Florida, 32127

GLOSSARY

AAA — Anti Aircraft Artillery
A/C — Aircraft Commander
A/3C — Airman 3rd Class
AD Raider — Aircraft Fighter Plane
AFB — Air Force Base
AP — Associated Press/Armor-Piercing Ammunition
APA — Naval Assault Transport
APO — Army Post Office
BAR — Browning Automatic Rifle
Bn — Battalion
CCF — Chinese Communist Forces
CCG — Combat Camera Group
CFC — Combined Forces Command
CD — Compact Disk
CO — Commanding Officer
CPL — Corporal
CQ — Charge of Quarters
DFC — Distinguished Flying Cross
DMZ — Demilitarized Zone
FO — Forward Observer
GI — General Infantry/Government Issue
GDP — Gross Domestic Product
GYSGT — Gunnery Sergeant
HQ-HQS-HQTRS — Headquarters
HVAR — High Velocity Aerial Rocket
KWVA — Korean War Veterans Association
LSO — Landing Signal Officer
LST — Landing Ship Transport
LT — Lieutenant
MASH — Mobile Army Surgical Hospital
MAG — Marine Air Guard
MCRD- Marine Corp Recruiting Depot
MIA — Missing In Action
MOS — Military Occupation Specialty
MLR - Main Line of Resistance
MP — Military Police

MRL — Multiple Rocket Launcher
MSR — Main Supply Route
MSGT — Master Sergeant
NAS — Naval Air Station
NATO — North Atlantic Treaty Organization
NCO — Non Commissioned Officer
OP — Outer Perimeter
PBY — Type of seaplane
PFC — Private First Class
POW — Prisoner of War
PUC — Presidential Unit Citation
PVA — Peoples Volunteer Army
PVT — Private
PX — Post Exchange
Q — Quonset Hut
RET — Retired
ROK — Republic Of Korea
R & R — Rest & Relaxation
SFC — Sergeant First Class
SGT — Sergeant
SSGT — Staff Sergeant
S.P. — Shore Patrol
TacAir — Tactical Aircraft — Protect Ground Troops
TAD — Temporary Assigned Duty
TDY — Temporary Duty
TSGT — Technical Sergeant
US — United States
USAF — United States Air Force
USMC — United States Marine Corps
USN — United States Navy
USNS — United States Naval Station
UN — United Nations
VHF — Very High Frequency
WAC — Women Army Corps
WIA — Wounded In Action
WWII — World War II

COLOPHON

THIS BOOK WAS PRODUCED USING THE FOLLOWING PROCESSES:

Research and gathering:
    Web: MS Explorer 6.0

Printer: Epson Stylus

Writing and manuscript building:
    Manuscript preparation: MS Word

Copyediting: Peggy Painter, M.Ed.

Cover Designer: John Morris-Reihl, Art and Technology,
    http://www.artntech.com

Image: Front — USMC; Back — *Time Magazine*

Design, typesetting & layout: CAI Publishing and Martha Williams Nichols, aMuse Productions®, *productionwoman@yahoo.com*

Typefaces:
    Body text: BarbedorTReg, 12 pt
    Running Heads: BarbedorTReg, 12 pt
    Chapter numbers: ITCBenguiatGothic Bold, 14 pt
    Chapter titles: ITCBenguiatGothic Bold, 21 pt
    Head 1: ITCBenguiatGothic Bold, 18 pt
    Head 2: ITCBenguiatGothic Bold, 14 pt
    Quotations: BarbedorTReg, 11 pt
    Quotes/News Stories: Times, 10 pt

Conversions: MS Word to QuarkXPress 4.1.1 to PDF using Adobe Acrobat 6.0

Printing: Lightning Source, Inc., La Vergne, TN; from PDF

Paper: 55lb offset, 444 PPI, white, acid free.

Cover: Four color, layflat film lamination; Verdana, Bold

Binding: Perfect bound (adhesive spine, soft-cover)

QUICK ORDER FORM

Email orders: info@caipublishing.net;   *Fax orders:* 1.440.306.0649

OR—Send this form.

Telephone orders: 1.386.383.5198.
Have this form and your credit card handy.

Postal orders: Send this form to CAI Publishing, 807 Black Duck Drive, Suite A, Port Orange, FL 32127-4726, USA. *Telephone:* 386.383.5198

Please send the following books.* I understand that I may return them for a full refund—for any reason, no questions asked.

_____
_____
_____

Send FREE information on:
[ ] Other books [ ] Speaking [ ] Coaching [ ] Consulting
Name: _____
Address: _____
City: _____ State: _____ Zip: _____
Telephone: _____
Email: _____

Sales tax: Please add sales tax at point of delivery for books shipped to Florida addresses.

Shipping and Handling: US: $6.00 for the first book or disk and $3.00 for each additional product. International: $15.00 for first book or disk and $7.50 for each additional product (estimate).

Payment: [ ] Check   [ ] Visa   [ ] Discover   [ ] Master Card
Card number: _____
Name on card: _____
Expiration Date: _____/_____

All credit cards will be processed through PayPal.

* See *http://www.caipublishing.net* for book selection

www.ingramcontent.com/pod-product-compliance
Lightning Source LLC
Chambersburg PA
CBHW071835230426
43671CB00012B/1965